Patrick M. Kiely

Southern Fruits and Vegetables for Northern Markets

Patrick M. Kiely

Southern Fruits and Vegetables for Northern Markets

ISBN/EAN: 9783337376987

Printed in Europe, USA, Canada, Australia, Japan

Cover: Foto ©berggeist007 / pixelio.de

More available books at **www.hansebooks.com**

SOUTHERN
FRUITS AND VEGETABLES
—FOR—
Northern Markets.

What to grow; how to ship and pack; the best varieties; the prices prevailing in St. Louis throughout the year,

And a variety of information of interest to Southern growers and shippers.

By P. M. KIELY,
March, 1888.   ST. LOUIS, MO.

This Pamphlet is free to all applicants who inclose three cents in stamps to prepay postage.

TO OUR PATRONS AND THE PUBLIC:

From every Section of the South we are constantly receiving letters in relation to the shiping of fruits and vegetables to this and other markets, the most profitable kinds to grow, how to pack and ship, the kind of packages required, the prices prevailing throughout the year for the various articles, and the great fund of information in connection with the trade, so important to everybody embarking in the business. As new parties are steadily going into business, we are besieged each season with the same questions. To supply this information, in a concise and convenient form, we have published this pamphlet, believing it covers most of the questions usually asked. We gather the information given from twenty years' experience in the business in this city, and we trust it will be of service to the many who will receive it.

P. M. KIELY & CO.

Spring, 1888.

JUST two years ago we published a pamphlet similar to the present work—printing an edition of 6,000 copies, and figured on that supplying the demand for three years. Upwards of 3,000 copies went out at once to our patrons and the various parties throughout the South and West seeking it. We reserved the remainder for new applicants, but the calls became so frequent, that we found the supply almost exhausted towards the 1st of February, 1888. After that time the demand became greater, until we finally concluded to publish the present volume, which we trust is an improvement on the old one. The many new parties going into the business throughout the South, coupled with the warm indorsements of the work by the newspapers and agricultural journals, has created a demand for it far beyond our expectations. The calls for it came from Florida, Louisiana, Mississippi, Texas, Georgia, Tennessee, Arkansas, Kentucky, Missouri, Illinois, and not a few from points further North and East, and from parties who wanted to embark in the business further South.

RETROSPECTIVE.

In connection with our subject, a few words concerning the growth and history of the trade in this city will not be out of place here. When the writer embarked in the business in this city, in the fall of 1866, there was but one fruit commission house in St. Louis, and, indeed, one house was all that was then necessary to take care of what was consigned here. All the Southern States, including Arkansas and Tennessee, were then unknown as shippers of fruits or vegetables. Southern Illinois was then the remotest point as a field for such supplies known to the city, and the fruit season was so short as to be of little value or interest. Since then a dozen or more of houses have sprung up, many of whom we believe claim now to be the oldest and most experienced in the city.

Each year new railroads opened up new fields and new territories, from which supplies began to come liberally, notably Arkansas, Tennessee, Mississippi and Louisiana, with shipments from more Southern points later. Each year the season lengthened, until the present time, when we have an unbroken selling season of nine months, beginning in March with strawberries, and ending about the 1st of December with grapes. The very extensive vegetable supply lengthens the season—in fact, keeps coming here throughout the year. An increase of commission houses, to take care of this constantly growing trade was, of course, a natural result, and tended to develop and encourage the production.

THE FUTURE OUTLOOK.

That there is a growing appreciation of fruit as an article of food, and very justly so, cannot be denied. The more fruit we consume the healthier we become as a people, and the less doctor bills we will have to pay. The fruit grower, in addition to being a public benefactor, finds some consolation in the fact that his calling, properly and intelligently pursued, is highly remunerative, paying much better in fact than numerous undertakings, claiming more public attention, in which considerable capital must be invested before anything can be realized. It is not as uncertain or full of the elements of risks as many other enterprises are, and, therefore, is a more inviting field for men of limited means.

The most encouraging feature in connection with the growing of fruits and vegetables is the rapid progress made towards utilizing the surplus. The past year or two introduced so many new canning establishments and evaporators and fruit dryers, and into so many districts where they were comparatively unknown, that a new hope has been inspired and a fresh impetus given to the business. Indeed the question of what shall we do with our surplus, need no longer fill the heart of the grower with dismay. The demand is becoming more general, not only for the fresh supply, but also for the canned goods—both fruits and vegetables, and the industry of growing and shipping has grown to such dimensions that a great variety of fruits and vegetables have become within the reach of all.

LOCATION—SHIPPING.

You should aim to get as near the depot or shipping point as possible. Long hauls, especially over rough roads—unpleasant features that many shippers cannot avoid—inflict on the fruit frequently very serious injury, especially if the art of packing for such emergencies is not thoroughly understood. You are too often in a hurry and your fruit is shook up, and you haven't time to examine it at the depot. Later, when the returns come in, if they do not compare favorably with your neighbor's, who placed his fruit in splendid order on the train before starting, the commission man " catches it," unless you devote a little time to reflection over the matter. If the receiver writes and explains, it may refresh your memory and make his offence less grevious, but if he does not do so, he will in most cases lose a customer.

You will not, of course, attempt to carry strawberries or other tender fruits and vegetables to town or depot in a wagon without springs, and your goods must be protected from the hot sun, the dust of the roads, and the rain, by a waterproof covering that will afford ample protection. Sufficient time must be had when loading up, to handle carefully, both at home and when the depot is reached. With these precautions properly observed, the prospects are that your fruits, etc., will reach the consignee in good order.

FREQUENT HANDLING FRUIT IS EXPOSED TO.

The average shipper has no idea how often his fruit is handled and moved about before it reaches the consumer, and therefore the importance of the most careful packing cannot be lost sight of. To illustrate, let us review the scene on the arrival of the fruit runs from the South—the two main runs arriving about same time in the morning (6:30 to 7:30).

On arrival of trains at the Union depot, the Southern & Pacific Express Companies back up their wagons to the express cars as soon as the doors are open. A few expressmen, assisted by some of the commission men, or their employes, enter the car and commence passing out the goods to the drivers. A dozen or more firms are represented, and all are in a hurry and anxious to get off with the fruit, for their customers are at their stores up town waiting, and they do not want to miss the early sales— always the best. Therefore, handling each package carefully or laying it down easily, is out of the question, where so many have to be handled in the very short time in which it has to be done. In this car is fruit from perhaps, 20 different shipping points, and from 100 different shippers, intended possibly for 75 different firms, for this car may have shipments for the various towns in Illinois, Indiana, Iowa, Missouri, Kansas, Nebraska, etc. The outgoing trains will soon be ready and all this fruit must be transferred, checked off and re-billed for its various destinations. The express employes, in their anxiety to keep these various lots from getting left, add to the confusion and prolong the delivery to local receivers. All must be separated for the various partties and numerous firms here and elsewhere. Consequently rapid and occasionally rough handling seems unavoidable. When the wagons are loaded they drive across the track to the express buildings and platforms, where the fruit is separated once more for the many firms whose wagons form a solid wreath around the platforms. It is lifted again and passed into all these wagons, receipted for and driven off rapidly, and on reaching the commission houses the fruit has to be separated once more and credited up to the respective owners and shippers. After being thus hurriedly handled half a dozen times it is ready for the purchaser's inspection. He throws it into his wagon once more with similar haste, and it is hurried off over the streets again, and set down again for the inspection of the consumer ; and it is safe to say, it could not be recognized now by the original owner—apart from his marks—unless the packing was of the best at the start.

These are some of the features of the business that should be calmly considered by the shipper, who, too often jumps at the conclusion that he was robbed—that his fruit was first-class, and must have opened up fine.

Large shipments or car load lots, do not, however, suffer to this extent, for such are usually loaded into the receiver's wagon and hauled direct to his store, or the Express Company's wagons will do the same when the amount reaches something near a load. Time and re-handling of fruit is thus saved to the large shipper.

IN REGARD TO PACKING.

Growers and shippers of fruit cannot realize, unless they were here to see it opened, how it injures the sale and depreciates the value of their goods to find inferior fruit mixed in, and covered up, in good fruit. Put in no inferior fruit of any kind. We know it is difficult to watch pickers where a great many are engaged, especially inexperienced hands, but the successful grower will take timely steps, whatever his hurry, to guard against such a serious mistake. Topping off, putting on top all the good fruit in the box is also a mistake, and its injustice must be apparent to the most indifferent. Let the surface represent a good average of the contents, or perhaps a trifle better, but no further effort should be made to practice a deception.

Remember your name or stencil number is on the packages, and the buyer commits to memory very readily the brand which deceived him.

Some of the crooked brands are so well known in this market that it is difficult to find a buyer for them, even at a big reduction. Every dealer is trying to secure the best trade which can only be accomplished by having nice, uniform fruit. We repeat, let your fruit run straight and do not injure your reputation by trying to deceive anybody. Packing is a most important part of the business and cannot be studied too closely, and you cannot get out of the business what it is capable of yielding unless your packing is done as it should be.

SOME FACTS TO CONSIDER

During the hot weather when you commence shipping. Peas and beans, for instance, gathered in the sun when the thermometer registers 90° in the shade, if packed immediately in a bushel box and put into the average hot car will soon be heated to 100 degrees, and a few hours later fermentation and decay follows. Moisture is the surest agent to hasten fermentation, decay and loss, and it is very important that the goods—whether fruits or vegetables—should be thoroughly dry, and the cooler you can get them the better the chances of their reaching their destination in good order. They often encounter while in transit, most unfavorable weather, such as close, cloudy, warm weather, accompanied by frequent showers, and unless the packing has been done under the most favorable conditions, goods will not arrive in good order under such circumstances. A most careful observer states that the crushed leaves of the radish furnish moisture enough to ruin the goods in 24 hours if packed in a temperature of 70 degrees or upwards. Exclude from the goods before packed, all the heat and moisture possible, and your packing shed should be so located and constructed that it will catch every passing breeze and allow the air to circulate freely, and thus carry off the surplus heat and moisture in the goods you are packing.

One error in packing, that is too frequently practiced is, that of putting into the same package the various grades, from green to ripe or over-ripe fruits, etc. If you will pack and ship either too ripe or too green—which we do not approve of—be sure to put them in separate boxes or packages, so that one will not spoil the appearance or sale of the other. You must remember that the inevitable jarring and jolting the fruit is subject to while *en route*, whether berries, tomatoes, peaches, or pears, will cause the hard ones to crush the soft ones, thus spreading the juice over all and spoiling the sale. You will therefore see the necessity of exercising proper precaution. Early in the season, when first shipments are made from the South, the weather is quite cool, and berries reach us as green as when they left shippers' hands, and do not ripen or color up on the way, but shippers' in their anxiety to catch high prices, pick and ship indiscriminately, and thus injure the market on themselves and their neighbors.

WHO TO SHIP TO.

To handle fruit to advantage requires experience and facilities which few commission houses possess. It can be readily seen that houses lacking experience, who receive such consignments only occasionally, are not prepared to do justice to shippers, or as well as those making a specialty of such products. A firm not regularly in this line of business sometimes receives a shipment when the market is weak and easily broken, and having no regular trade, is compelled to sell under the market price, thus precipitating a general decline, which could be avoided had the goods been held by some house having an established trade, We are not only familiar with the wants of the local trade, but have built up a good order trade, and at times use the wires freely, at our own expense, in the interest of shipper and purchaser, and thereby enable ourselves to clean up and save the market. Through our exertions, in this and other ways, we have done much to bring buyers as well as shippers to our market, and at the same time built up for ourselves a good business, our efforts being properly appreciated by all parties concerned.

We have no doubt the same remarks will apply with equal force to the leading houses in the same line in other markets, a number of whom will be found in this book, for we have selected the best and most experienced firms we could find in the markets represented. All the cards found elsewhere, form a part of the information which should go out with such a work as this.

SPECIAL NOTICE.

We wish to state to shippers, especially to the many new ones embarking in the business, that the prices received for Southern fruits and vegetables in the principal markets of the country, during the shipping season of 1887, were far above the average figures, the result of unfavorable weather—first, late frosts, and later, prolonged drouths.

Therefore, the prices we shall quote as prevailing at the respective dates given, must not be relied on as a fair average, or as a basis for future operations. A liberal margin must be allowed to strike an average.

At this writing (March 1st) we are assured that the berry crop of Louisiana is the largest she ever raised. The vegetable supply of Mississippi promises to be more than double that of last year. The strawberry crop further North, however, especially at the big shipping points in Tennessee, Arkansas, Illinois and Missouri, will be small, even below the light yield of 1887, the result of the prolonged drouths of the Summer and Fall.

FRUITS.

STRAWBERRIES

Are the first fruits of the season. They come to us with the genial atmosphere of Spring, though not infrequently with the raw winds of March, and occasionally with the snowflakes and hard freezing of February. However, regardless of the weather, they are warmly welcomed by the epicure, the invalid and by more or less people with fat pocket-books. Eighteen or twenty years ago strawberries in this market were something of a luxury. The season then was about six weeks in duration. Now, it is six months from the first receipts from the far South until the final shipment from Northern Illinois, or Racine, Wis. The strawberry has been steadily gaining in popularity. It merits the patronage of every man, woman and child. No healthier fruit can be consumed. It is eminently the fruit for the million and now so extensively cultivated, that it is within the reach of all. The supply, rapidly as it has grown, has hardly kept pace with the demand. The many new railroads penetrating every section has become the most important factor in the development of the business North and South, and served to bring together, in every market, both the consumer and producer. The re-shipments from here of the Southern products are very large compared to what they were a few years ago. St.

Louis is rapidly becoming, in consequence, a great distributing center, and now has access to a wide range of territory from which she was formerly shut out. Similar progress in the same direction has doubtless been made by other leading centers.

More money has been made off the strawberry than any other fruit, considering the time, labor and money involved, and it is likely to remain the most profitable. It represents more money to the acre, as well as more real profits, than any other product. Many of the Southern cultivators in the various states growing considerable small fruits are novices in the business, and have had to battle with all the obstacles and disappointments that beset the pathway of the inexperienced growers, and not a few became discouraged and dropped out when successful results were almost discernible.

It proved a great relief to many of them to be relieved from the unprofitable labor of cotton raising on lands eminently adapted to fruit growing, and yet, rather unproductive and unfit for Southern staple products. There is still a great deal of such land, largely impoverished by the incessant strain of crop raising, that the fruit grower could render remunerative by diligent effort.

The strawberry crop rarely fails and never proves a total failure, as many other fruit crops do, except through gross negligence. Take this county for instance, where the business is conducted very extensively, some 1,500 acres, and no such thing as a failure of the crop has been recorded in the past twenty years. Occasionally, the crop is light through most unfavorable weather or other causes, but half a crop is the lowest estimate that can be recalled since the business began here, in a small way, twenty years ago. You will see then, that the strawberry growers' investment cannot be regarded in jeopardy, as investments are in most other avenues of trade; and, while there is not the alluring profits in the business that there was eight or ten years ago, it must not be forgotten that the margins or profits in every line of business have not only declined and shrunk just as rapidly but to a greater extent.

The first receipts in our market of late years except irregular shipments from Florida, came from Louisiana, generally from Independence, Amite City, or Tickfaw, little stations not far north of New Orleans. The Crescent city, however, receives berries from her suburban fruit patches long before any city further north is favored with shipments, mainly because the favorite varieties there are too sensitive to stand long shipments successfully. Ten or twelve years ago, Citronelle, Ala., thirty miles norty of Mobile, furnished us the first berries of the season for several years in succession and was followed by Charleston, S. C., for a few years. Later, Texas had the honor of doing so for two or three years, and Mississippi came very near carrying off the distinction several times.

The first receipts of strawberries, last Spring, were from Florida, several cases on the 2nd of March, and sold from 75c to $1.00 a quart. The receipts continued from Florida for several days thereafter, but sales were slow at $3.00 a gallon. On the 4th some small lots came from Louisiana, but were rather green and poor, and sold lower than the Florida receipts—$2.00 to $2.50 a gallon. March 5th to 10th, under pressure of heavy receipts from Florida and most unfavorable weather, prices declined to $1.25 to 1.50 a gallon. From the 15th to 25th, the figures remained the same, Louisiana, Florida, Mississippi and Alabama now supplying the demand. About the 1st of April the market is $1.00 to $1.50 a gallon according to quality and condition, and the above four States, with a few small shipments from Texas, are represented. The first week in April, Mississippi berries are quoted at $11.00 a case, six gallons. On the 12th of April first berries from Arkansas appear, Crystal City variety, and sold $7.00 to $8.00 a case. April 15th to 20th reads: Mississippi receipts, $8.00 to $10.00 a case; Arkansas, $8.00 to $10.00, Scarlets, $5.00 to $6.00; Louisiana, $8.00 to $9.00; Alabama, $7.00 to $9.00; Florida receipts soft, and she disappears about this time. From the 23rd to 30th of April, most of the receipts are from Arkansas; Tennessee, however, has made a few shipments, and on the 1st of May receipts are quite heavy; Mississippi is still shipping—fruit is firm and selling at $3.00 to $4.00 a case; Kentucky appears 1st of May, her fruit selling at $6.00 a case; Arkansas, $3.00 to $5.00; Tennessee, $4.50 to $5.00 a case. 2nd and 4th, receipts very heavy—bulk from Arkansas—which sold at an average of $2.50 to $3.00 a case, fruit being delayed and some soft and water-soaked; Tennessee, $3.00 to $3.25; Kentucky, $4.00 to $4.50; Southern Illinois, $4.00 to $4.50; Southeast Missouri stock commenced to come, and sold at $4.00 to $4.50 a case. May 10th, quotations: Arkansas, $2.00 a case; Tennessee; $2.00 to $2.50, mostly Crescents; Kentucky, $2.75 to

$3.75, Southern Illinois and Southeast Missouri $3 to $4.00 a case. May 15th the Price Current reads Tennessee and Arkansas shipping in a small way, but fruit soft and inferior (final shipments). Arkansas $1.75 to $2.75, Tennessee $2 to $3.00, Southern Illinois $2 50 to $3 75, Southeast Missouri, mainly Monarchs, $3 to $4 25, Kentucky Crescent and Sharpless $3.00 to $4.25. Home-grown have now commenced to come regularly—75c. to $1.00 a gallon very fine Wilsons. After this time the market possesses little interest for the shipper south of this. It will be seen from the foregoing that prices did not reach the low point they often do at the height of the season. It should be added that an unfavorable season lowered the grade, or quality of the fruit for most shippers, especially those of Tennessee and Arkansas, which embraced for ten days or more the bulk of the receipts.

You will see from the foregoing where the fruit comes from, when they begin, and who you will have to compete with as shippers as the season progresses.

Our local crop of berries, usually very fine, is composed largely of Wilsons, and is shipped freely in every direction. It comes in half-bushel drawers, in stands of four drawers. Some, however, continue to use the six-gallon case, which many shippers prefer to any other package in filling orders.

As to varieties, we still consider the Wilson the great berry for commercial purposes. For the family wants or local trade we would select a sweeter and finer flavored fruit. The Sharpless, Monarch, Capt. Jack, Crescent, and many others we might enumerate, have claims to distinction too. The Charleston variety we regard quite valuable for the far South, say Central Mississippi and further South. A great many newer varieties than the foregoing are very highly spoken of, but not being very familiar with them shall make no further reference to them.

Six-gallon cases (24 quarts) bring most of the fruit to this market and will doubtless continue to do so. For long distances or Eastern markets would recommend the Gift Crate, a 32-quart ventilated crate (basket quarts), such as Florida uses, a package which meets all the requirements of thorough ventilation. The special paragraph elsewhere in regard to packing should not be overlooked.

BLACKBERRIES.

Do not figure very extensively among the shipments from the South. It is not a good shipper, and a good portion of the consignments arrive in bad order. Blackberries, under certain conditions, become sour while in transit during the night—though we have seen berries out thirty-six to forty hours which arrived in fair order. The very hot weather that usually accompanies the maturing of this fruit is the worst feature it has to contend with. We would not, therefore, advise extensive planting by parties far away from market. Arriving as it always does when the market is crowded with strawberries, it does not bring any fancy price. The fruit, however, is attractive and sells readily, if it can be placed before the purchaser fresh from the vines or a few hours after gathering.

The first receipts were wild from Arkansas, May 27th, and sold at $3.00 a case—six gallons. They continued to come steadily for ten days, the price varying but little, $2.50 to $3.00 a case. About the 10th of June cultivated are coming and find buyers at $3.50 to $4.50 a case. From the 15th to 20th receipts are very liberal—mainly wild—Arkansas still furnishing the majority of the receipts, although Illinois and Missouri are also shipping, and prices range from $1.25 to $2.50 a case for wild, according to condition, etc.; cultivated $2.50 to $3.50. The course of the market later will disclose little of interest to Southern shippers.

Alabama shipped here very successfully three years ago for nearly a month—commencing about the 8th of May and securing good prices $3.00 to $4.50 a case. It was a small berry, probably wild, but firm, and reached here by fast freight, two and a half days out, in fine order.

The Lawton—an old standard variety, one of the best, its only fault that it is a little tender and gets killed too often in this latitude. The Kittatinney is in a measure taking its place, being quite hardy and very productive, but the fruit is not so large. There are several other highly recommended varieties not so well known to us.

The strawberry case is the most suitable package. They should be gathered just as soon as fairly colored—while yet firm. If permitted to get fully ripe, or soft, will soon sour, the least jarring, or rough handling scattering the juice, which sours the whole lot in a few hours.

BLACK RASPBERRIES

Have been slowly but steadily disappearing from our market. The supply of last year was the lightest in ten years, though nine or ten years ago immense quantities came in from this vicinity—especially from Collinsville, Ills., ten miles east of us. At present, however, the people there are devoting their attention to something else, for the strawberries became so abundant and cheap in the market of late years that the profitable opening for black raspberries vanished.

Still there is too little cultivated now for the demand. During the past two years the shipper fortunate enough to have some realized good prices, and for the first time in the experience of the trade they sold as high as the red varieties, which formerly averaged nearly double the prices paid for black. The black is a good shipper, hardy and productive, and not as sensitive or difficult to grow as the red. For long shipments the pint box and three-gallon cases ought to be used, though parties within a few hours' ride of the market, could use the quart boxes and crates. They appear soon after the blackberry, and open at about 75c. a gallon, and gradually decline as the receipts increase, until they strike $2.50 a case (24 quarts), which is about as low as they reach at any time.

There is a good demand now for the black as well as the red raspberry, from the canning and preserving establishments, an additional inducement to cultivate them. We consider the MIAMI BLACK CAP and the GREGG the two best varieties grown—the GREGG heading the list.

RED RASPBERRIES

Are a prime favorite with all lovers of fruit, and yet have been somewhat overlooked by the average fruit grower until about four years ago, when a number of cultivators turned their attention to them. Southern Illinois grows them in abundance and ships freely to this city—always in pint boxes, in neat, flat, three-gallon cases, which are the proper packages for this delicious, but somewhat delicate fruit. They do not stand long shipments. Arkansas or West Tennessee is as far south as we would suggest growing for this market. Mississippi grows considerable of them for the New Orleans and other Southern markets, and profitably too, I learn. When the receipts become large, the canning establishments here are the most liberal buyers we have—their figures, according to supplies on market, quality and condition of fruits, are from 40c. to 60c a gallon, but this is in the midst of the season, when they are most abundant, say from the middle to the latter part of June. The first receipts came in 25th May from Arkansas, and ranged from $2.00 to $2.50 per three-gallon case, pint boxes, and slowly declined until the latter part of June, when they fell to nearly half the foregoing figures, the lowest prices known to the trade here. The prices about 1st July are $1.00 to $1.50 per three gallon case (24 pints), and the receipts mainly from Kentucky, Southeast Missouri, Southern Illinois and this country, ranging in quality and condition from very poor to choice.

CHERRIES

Do not appear to be a very profitable crop south of this latitude. The cherry tree being perfectly hardy, thrives in all the Northern States in good soil, and it is not adapted to a warm climate. The finer varieties which embrace the Mazzards, Hearts and Biggareaus, do not flourish in either the West or South, owing principally to the injury inflicted on the bark or trunks of the trees by the hot sun of midsummer. The Dukes and Morelloes are less susceptible to climate influences, are smaller and hardier, and the fruit being more acid, embrace some of the features that make it a better keeper and shipper hence they are better adapted to the West and South. Illinois and Missouri furnish most of the cherries consumed here.

The first receipts received were from Southern Illinois, 14th of May, and sold at 75c. per gallon—Early May variety. The first receipts the previous year were from Arkansas, and sold at 65c. per gallon, or $4.00 per case, getting in on the 15th of May. In 1886, first receipts were from Tennessee, 18th of May, selling at 75c. per gallon. The prices do not decline much for the two weeks following, but then the Illinois and Missouri shippers commence and ship quite freely, and the price soon declines to $2.00 per case, and later, when the growers in this vicinity get to picking, the price is down to 25c. per gallon, which is generally the lowest they reach at any time, figures that shut out growers at a distance. In damp, cloudy weather they decay very rapidly after reaching maturity.

FOR NORTHERN MARKETS. 23

Of course the stem should be left attached to the fruit, since it not only fills the box much more readily in this way, but keeps the berries from bleeding and becoming sour soon afterwards. Cherries come to us occasionally from as far south as Mississippi. Every fruit grower should have at least a few trees for home use if nothing more. The strawberry boxes and crates should be used for them.

GOOSEBERRIES AND CURRANTS

Receive little attention at the hands of Southern shippers. The climate is not so well adapted to their culture as it is further north. Some have tried them, no doubt, but with what success we have not learned. We consider each worthy of trial to some extent, at least, for local consumption if nothing more. It would pay well to get either in here ahead of local growers. They are not grown extensively in this section, though the prices are usually good throughout the season.

Our market is supplied with currants mainly from Northern Illinois. Onarga is famous for currant growing. Galena also ships considerable to this market. Strawberry boxes and crates are adapted to both in shipping.

Gooseberries were quite scarce in this market last year, $2.00 per bushel being the lowest price paid, while the bulk of the sales were $2.00 to $2.50 per case—24 quarts. Supplies were evidently small at all points. *Houghton, Seedling* and *Downing* are the best varieties. They heat in a short time in barrels or sacks, and should be shipped only in drawers or strawberry cases—24 quarts. They stand shipping so well is one good inducement to plant. Ship when full-grown before they commence getting brown.

CURRANTS flourish best in a cool, shady or partly shaded locality, such as northside of fences. They grow successfully in this locality and ought to succeed further south. The Red and White Dutch varieties are best. Use the same packages as for gooseberries, drawers or quart boxes. They were scarce and high all last season—ranging from 40c. to 75c. per gallon throughout.

NECTARINES

Are entirely neglected or forgotten by the fruit growers patronizing this market. Indeed, they are a novelty here, so rarely can they be found. When they do appear, however, they find ready sale at $1.00 to $2.00 per box. The Nectarine is simply a peach with a smooth, glossy skin, devoid of the fuzz of the peach, but its smooth surface unfortunately seems to attract the attention of the Curculio who preys on it, and is, no doubt, largely responsible for its absence in our markets. We believe, however, it has not been given a fair show by the fruit growers, or we would see more of the fruit.

APRICOTS

Ripen a month before peaches do, and would strike a splendid market on this account, and nearly all I have said about the Nectarine will apply with equal force here. It is too much neglected and its great enemy, too, is the Curculio, which attacks the plum, a fruit the Apricot so much resembles, partaking of its character and habits, and successfully attacked by the same insects. The Apricot is budded on seedling Apricots, and also on peach and plum stocks, the latter preferable, being longer-lived. The Apricot appears a cross between the plum and the peach, but from a scientific point, it is not. A few come in from this county and find willing buyers at $1.50 to $2.00 per box, but very seldom, however, can any be found here.

WHORTLEBERRIES

Come to us quite freely every year, especially from Arkansas, where they grow wild. They usually sell well at $3.00 to $4.00 per case of twenty-four quarts. A good many are sent in only partly ripe, and often mixed, some green and some ripe. Green ones are unsalable and have to be dumped, while the mixed sell according to amount of ripe ones in the package. Only straight ripe or fully colored should be shipped, as it is difficult to sell the mixed or partly ripe at any price.

THE DAMSON PLUM

We believe, has never been properly tested by the fruit growers generally of the South, and I doubt whether any other plum will pay near as well in the territory tributary to this market. Very hardy and productive and enjoying, as it does, immunity from insect enemies, and in addition to these market advantages, the best shipper of all—it is more than surprising that it should be so overlooked. It thrives on neglect, yields a half to a full crop as regularly and surely as the apple orchard does, and being long-lived and content to flourish in out-of-the-way places and fence corners, it appeals strongly to the fruit grower for recognition. The market moreover is never glutted, rarely sells below 75c. to $1.00 per box, and more frequently averages $1.00 to $1.25 per box, and comes through in splendid order in one-third bushel boxes. It should be gathered when fully colored, and will then be safe for several days' shipment. It also makes a most delicious preserve, and is purchased freely for this purpose by the many preserving establishments here who have to send East frequently for supplies.

Very few of the choice varieties of plums, so well known East, are grown anywhere within reach of this market. The many who attempted cultivating an assortment of choice varieties became discouraged at the inroads made on the crop by the Curculio and other insect enemies, and did not exercise the patience, perseverance and labor necessary to save the fruit from them, so abandoned further efforts in that direction. Very rarely, therefore, can the Gages or other favorite varieties be found in our market.

QUINCES.

There is little demand in this market for the quince, until the heated term is passed. Being purchased only for preserving purposes, they are somewhat neglected until toward the 1st of October. Most of the preserving is attended to in October and November. This suits the producers within a radius of 100 miles or so of this market, but the more southern territory find this too late for their shipments, which mature some weeks earlier. Prices during August and September average $1.50 to $2.00 per bushel, and for the next two months $2.00 to $2.50 per bushel. If the fruit is gathered carefully and kept entirely free from bruises, and laid away in the coolest places accessible to the grower, in the absence of cold storage, they can be kept successfully for several weeks. Quinces come here every year from California, wrapped in paper in three-peck boxes, but do not reach here until the local supplies disappear. As few fruit growers pay any attention to the quince, it is usually a profitable crop.

The majority of the quinces that appear in our market every season, usually in November, come from New York State, where they are extensively and successfully grown. In the West and South, the few trees planted have been too much neglected, and as a result, many of the trees become stunted and barren.

The soil for the quince should be deep and rich, such as will raise good corn and potatoes, and should be kept well cultivated. I have rarely seen a quince tree in my travels that was not stunted and full of suckers, the usual evidence of neglect.

They can be packed in half, or bushel boxes, also in barrels, and can be shipped by freight when the express charges come too high. When full grown, but before they color up much, gather and ship them. This will afford an opportunity to re-ship if necessary.

GRAPES.

Grape growing south of this latitude has not received the attention that the business merits. Instead of increasing, the business has been on the decline for several years. Those that have fairly tried it to an extent, have made it pay handsomely. From what we have received from the different States south of this latitude, we think that the Ives Seedling, Concord and Delaware are the varieties that will produce the most money. Of course, several more varieties might be profitably grown, but we would head the list with these three varieties, and next, would say Moore's Early, similar to the Concord, ten days earlier, and quite valuable on this account. We disclaim any intention of doing injustice to the many newer varieties that are offered, some of which may possibly prove more profitable than the varieties well known to us, so the matter of testing, etc., rests with the grower. The experiments made are few, and possibly the grape for the million in the Southern States is still unknown to them. The Scuppernong, so well known to the Southern people, is unknown in this market, rarely appearing here; it would not become popular here, anyhow. The Concord is here among grapes what the Wilson is among the strawberry buyers, the one for the people, especially for commercial purposes. We would suggest discarding the Hartford altogether. The berries drop off when it reaches maturity, and it becomes at once almost unsalable, unless for a mere trifle.

The injury inflicted on the crop in this vicinity by late frosts and severity of midwinter, the past three or four seasons, discouraged so many grape growers that we look for little from them in the future. In fact, within a radius of one hundred miles of this city the business is steadily on the decline, and far from what it was ten or twelve years ago, when Missouri promised to become a great grape growing State, and when a number of confident and enthusiastic growers were making extensive arrangements for the future. In consequence, last year St. Louis received from Ohio alone, during the six weeks she was shipping, thirty-five to forty cars—25,000 pounds to the car—the majority of them in neat ten-pound baskets, wooden covers, the best package used. New York State shipped here also, perhaps fifty cars in all, the bulk of them choice Concords. Pennsylvania and Northern Illinois, also shipped several cars. A good many of the Ohio receipts were also in open baskets, holding about fifteen pounds; a brown paper over them and tied around the basket with a string. On arrival the papers were removed, and the bunches of grapes which rounded the top were so carefully handled that the bloom on them remained undisturbed. They looked inviting and sold well. This package can be used only when shipping by the car which is locked, sealed and open only by the consignee at destination. Among the New York and Ohio grapes last season were some of the Niagara and Pocklington varieties, the finest white grapes that ever appeared on our market. The bunches are large and compact—berries as large as the Concord—good flavor and shipper, and altogether highly attractive in appearance. They sold readily at 10 to 12 cents per pound when the market was crowded with grapes. Ohio grows the Concord and Catawba very successfully, and the art of handling and packing is thoroughly understood by the growers there. Western New York shipped very choice grapes—the handling and packing being about perfect. They came during October and November. From the foregoing it can be readily seen that there is more encouragement than ever for Southern growers, as the markets of the North and West will be found comparatively bare for a long time.

The first receipts last year appeared the first week in July, were from Mississippi and Alabama, one-third bushel boxes, 12 to 14 pounds net, and sold at $1.00 per box. The variety was the Hartford—the poorest variety cultivated, having really nothing to recommend it. Any other variety coming at this time would sell much higher. Texas made a few shipments about same time, crowding the fruit into quart boxes in 6-gallon cases—a very poor package to use or sell.

A week later, July 15th, Hartford's were quoted at 5c. per pound; Ives, 7 to 8c.; Concords, first receipts, 10c.—the majority of the receipts in one-third bushel boxes. Delawares

FOR NORTHERN MARKETS. 31

appear about this time and sell at 12c. per pound. On the 1st of August, we find Delawares selling at 10 to 13c. per pound; Hartford's, 4c.; Ives, 6 to 7c.; Concords, 7 to 8c.; Martha, 6 to 8c., and the receipts are mainly in ten pound baskets, a much more desirable package, but selling on the above basis and gross weight. At this time and later on, all kinds of packages are coming. About the 15th of August, when the market is lowest of the season, broken down rather by the poor quality than quantity coming, mainly local growth, the quotations are: Hartford, 2c.; Ives, 3c.; Concords, 3 to 5c.; Delaware, 10 to 12c.; Martha, 5 to 6c.; Goethe, 6 to 8c.; Elvira, 7 to 9c. The markets remain about this way a week or so, and after that time begins to improve. The Virginia Seedling is the last variety to appear here—selling mainly to the wine-makers at 4 to 5c. per pound. When the local supply is nearly exhausted, Ohio begins, and New York soon afterwards. The receipts from this vicinity and southward last year, disclosed the poorest packing we have witnessed for years. They were cut off the vines and packed regardless of their condition. Now permit us to say, only fully colored or ripe grapes should be shipped. Cut off the green, rotten, shriveled, dried, or otherwise imperfect fruit. All should be cut out carefully with a pair of scissors before packed. The regular ten-pound basket, that used by the New York and Ohio grape growers, is the package that should be universally used.

To Southern shippers we will say that grapes come in good order also, in one-third bushel boxes, when properly packed. If loosely packed, or in such a manner that any of the grapes can be displaced or moved while in transit, they will not reach here in good shape. The bunches should be laid in carefully, in rows like peaches. At the top let the cover press down sufficiently to hold firmly all the bunches in their places. When the cover is removed on arrival here no stems should be in sight, only a smooth surface of grapes should appear. In packing let the stems be downward. The fruit should be handled as little as possible, so as to protect the bloom that covers the grape. Packing in quart-boxes, in six-gallon cases, should be avoided. The fruit has to be handled too much, and the bunches are not such size as will fill the boxes to advantage. Five to ten-pound boxes in crates or frames might also be used, but the basket is the proper package for the grape, and early in the season, long before needed, correspond with some leading establishment, such as the one at Cobden, Illinois, whose card appears elsewhere, and ascertain the kinds offered and costs of same. We repeat, you have a long and profitable season in all the Western markets before you are disturbed by competition. The keeping qualities of the grape is one of the important features to consider. The Ohio and New York grape shippers can hold their grapes in buildings prepared for that purposes—cold storage apartments, etc.—three to six weeks, or until a better market appears, and then ship when the best prices prevail. We have Catawba grapes in this market to-day, February 28th.

PEACHES

Are becoming more important each year as a crop to Southern fruit growers. Thousands of acres in most of the Southern States, more especially in Arkansas, Tennessee, Mississippi, Northern Louisiana and Northern Texas, are eminently adapted to peach growing.

The business has undergone a revolution the past few years. Five or six years ago the culture of early sorts only was considered south of this latitude, especially in Arkansas and Tennessee, and many large orchards were planted about that time; the obnoxious and unprofitable Hale's Early figuring extensively in the selected list. Three and four years ago the earliest sorts not only failed to pay, but in many cases it would have paid the grower to let them rot on the trees or under them. Still the shippers thought they would give them another chance, but experience with these earliest varieties three years ago was bad enough to condemn every tree. It is safe to say, that more than half the shipments were consumed by the express charges. A rooting out of these very early sorts followed to a beneficial extent, and the past two seasons showed the wisdom of the step, in the lighter receipts and better prices resulting.

It will be seen, then, that the former plan will have to be reversed, as it is the medium to late varieties that pay best now. The Troth's Early, or the season of its ripening, is early enough, and those varieties maturing before that, rarely pay. They are entirely too perishable in their character, and warm, rainy, cloudy weather affects their appearance in a few hours, and the shaking up they get by the many handlings they are necessarily subject to, assists in rendering them unsightly if not unsalable in a short time.

A few years ago, when Missouri and Illinois were growing far more peaches than they are now, they were the great competitors of the more southern shippers, and the Arkansas, Texas, Tennessee and Kentucky growers realized that their only hope was in early shipments—since the advantages of the season, or earliness in maturing, gave them the field for at least several weeks, and, hence, the great fields of early sorts. Now that Missouri has had five successive failures of the crop, and Illinois only two small crops in five years, they are almost out of the race, for not only have the buds been killed each year, but many of the trees were ruined also by the severity of the weather, and very few orchards or trees were planted to take their place. The outlook, therefore, is very bright for the more southern shippers; they must abandon the idea of growing the earliest kinds, however.

In regard to varieties it is very difficult to arrange a list to suit such a wide range of territory, so we will not attempt to name one.. A list in Southern Illinois may not meet the approval of a Texas or Mississippi grower, and Arkansas might select a list differing from that needed in any other State. There is one variety that has found general favor in the South, and that is the Chinese Cling. We think it has received more attention than it merits. It generally lacks color and is too subject to decay—serious objections. The bulk of the receipts are of a greenish color. They have size, a desirable feature, but that is about the best thing that can be said for them. The finest fruit of this variety comes from Texas, being not only large, but finely colored and generally sound, and bring a fine price.

The following list embraces ten first-rate varieties, ripening in this vicinity, in the order named, from first of August to end of October,—nearly three months. A majority of them mature in September. *Mountain Rose, Crawford's Late, Reeve's Favorite, Freeman's Late, Old Mixon Free, Lemon Cling, Susquehanna, Stump the World, Ward's Late, Heath Cling.*

The first receipts of the season last year arrived on the 20th of May from Texas, and sold from $2.00 to $3.00 per box, according to quality—regular peach boxes, one-third bushel. Receipts continued light for several days thereafter, and a good many of them too green and poor to sell to advantage. On the 25th to 30th of May, more liberal receipts, embracing from fair to poor, green, etc., and the figures 50c. to $1.50 per box—a few going for express charges—hard, green, specked, etc. May 28th, Arkansas receipts are quoted, the prices a trifle higher than the average Texas receipts. The first week in June, a good many are coming, "it would pay better to keep at home." The prices paid are 25c. to $1.00 per box, bulk sales 25c to 50c., receipts mainly from Arkansas and Texas—Mississippi shipping a few. June 6th to 12th. The most of the fruit is coming from Arkansas. Considerable of it too green and small, and the figures 50c. to 80c. per box. Texas is shipping some very poor fruit at this time, also small, green and hard, and averaging but little above the express charges—too many specked ones being among the receipts—the *Hale's Early* being in especially bad condition through decay, etc. The 10th to the 12th, the first receipts from Southern Illinois, Tennessee, Kentucky, and Southeast Missouri, are noted, and prices 50c. to $1.00 per box, while the Crawford's have just appeared from Texas and are selling at $1.25 to $1.50 per box, and eagerly sought for.

The 15th to 20th of June, Illinois, Kentucky and Missouri receipts quoted 50c. to 75c. per box. Arkansas and Tennessee about same price, while Texas with later and better fruit, is quoted 60c. to $1.50 per box, according to quality, condition, etc.

For the next ten days there is scarcely any change to note. July 1st, the market quotations are: Texas, 50c. to $1.35; Illinois, Missouri, Kentucky, Arkansas and Tennessee, from 50c. to 75c. per box. A good many peck boxes coming from Texas, and proportionately lower than the foregoing. First week in July, weather hot and unfavorable, and considerable Texas stock arriving specked, and selling at 60c. to 75c. per box for the best varieties. July 10th to 20th, Texas peaches sold at 50c. to $1.25; a majority of them *clings*, and selling from 10c. to 20c. per box lower than the freestones. The receipts for Illinois, Missouri, Kentucky, Tennessee and Arkansas, 50c. to 90c. per box. July 25th to August 1st, Texas clings sold at 50c. to 85c.; freestones, 65c. to $1.10. Arkansas, Tennessee, Illinois and Missouri, 50c. to 80c. per box. August 1st to 10th, Texas stock falling off; very little coming after this time—prices mainly 60c. to $1.00 per box, but later and better varieties are coming from other States, and prices are 50c. to $1.00 per box.

After the middle of August, the bulk of the receipts are from Illinois. Nice stock generally, and prices are 75c. to $1.00 per box up to 1st of September, when Michigan commences, with nice fruit, packed in very neat peck baskets, which finds ready sale at 50c. to 75c. per basket. During September, Illinois and Michigan furnish most of the offerings, and the Southern shipper is no longer interested in the course of the market, but prices gradually improve after the above dates. During the season Northern Louisiana made several shipments, but generally did better in the large Texas towns—so much nearer home. The Chinese cling came regularly from the South, but Texas, where it *colors up finely*, is the only State where we would advocate its cultivation.

One-third bushel boxes should always be used. In packing, set the box on the edge (not on the flat), place the fruit in rows along the edge of the box and fill up carefully. Let the cover press lightly on the fruit in nailing it on. When the package is in good shipping order, unbroken rows will appear at all the openings and not a peach can move. We invite the attention of our readers once more to remarks on packing fruit.

All fruits packed in one-third bushel boxes should be handled and packed in the same manner as peaches, always setting the box on the edge when starting to fill.

PEARS.

The pear crop of the West and South was small the past two years. We had to look Eastward for part of our supplies. Western New York, the greatest apple and pear growing region in the United States, shipped pears by the car load to Western markets, many of them going to Chicago, Cleveland, Cincinnati, St. Louis, Kansas City, Omaha, Denver, etc. In every shipment the Bartlett predominated, being a prime favorite in all the Northern and Western markets. It must be admitted, however, that the pear crop of the South was very light, at least, this market remained almost barren of the Southern product. The Illinois and the Missouri yield was quite liberal during August and September, and prices went lower than the previous years. I do not believe that the pear in the West and South has had yet a fair chance to show what it is capable of yielding in the way of profits to the cultivator. The greatest trouble is the little care or labor bestowed on the orchards, or few trees planted. *Neglect* is the main cause why Western markets are so poorly supplied with the native or local growth. Bartletts, Duchesse, Seckel, Louis Bonne, White Doyenne, Clapp's Favorite, Flemish Beauty, Lawrence, etc., are the best known varieties in this market.

The LeConte, the new favorite in the South, has been more widely discussed than any new variety that has appeared the past fifteen years. It appears to flourish in the Southern States, almost, if not quite, free from the Blight, a most important advantage. Here is how J. J. Thomas, very eminent authority, describes it: "Large, yellow, moderate in quality; exc edingly productive, vigorous and productive at the South; of no value North. Season midsummer." The Kieffer, another new candidate, has received a great deal of public attention the past few years, but it is generally admitted that in quality it is far inferior to the LeConte. Here is the description: "Rather large, oval, contracted towards stem and crown; rich yellow, tinged with red; flesh, varying from coarse to fine, and from good to quite poor in quality; good for canning; tree vigorous and very productive." The Kieffer, it is claimed, is remarkably free also from the Blight, the great enemy of most pear orchards. We received some of the LeConte from Texas last summer. They came nicely packed in peck boxes; fruit large, greenish, not colored, having been gathered a little too soon. It sold readily at $1.00 per box or $4.00 per bushel, a good price, condition considered.

The first receipts were from Mississippi, about the 15th June, and sold at $1.25 to $1.50 a box, and for two weeks afterwards receipts were quite meager and prices $1.00 to $1.25 a box. About 1st July home-grown appeared and sold at $2.00 to $3.00 a bushel. Receipts continued light thereafter for two weeks and prices declined but very little, except for the small hard sugar pear, which sold at 50c. to 65c. a box. August 1st, Bartletts are quoted 70c. to 80c. a box and other varieties 40c. to 65c. according to quality. The first fifteen days in August showed lower prices by ten cents a box. Early in September Eastern receipts appeared by the car load and, coupled with home-grown crop, caused the lowest prices of the season. Bartlett being quoted at 50c. to 60c. a box and other varieties 40c. to 50c.—the latter for Duchesse. At this time, or rather from middle of August to middle of September, cold storage is the place to put pears, for October always shows a big advance—nearly double the prices of a month previous. Several parties both in Illinois and Missouri stored very profitably about that time last year. The cost of doing so—60c. to 75c. a barrel—is not much compared to the advantages derived. Should be stored only in barrels, and full ripe fruit should not be packed, as it will not keep, but injure the others while stored.

The pear is a rich, luxurious fruit when grown to perfection or properly matured, and is marked for its great delicacy, juicy textures and delightful flavor. It is, of course, a favorite with the public for its many excellent qualities. The pear properly managed is an excellent shipper. It should be packed in the regular peach box, but when the business is conducted on a more extensive scale, as it is in the East, the barrel is the most economical for general use. They should be gathered when full grown, but before they are fully colored. Do not wait till they get mellow, as that desirable condition will be reached while the fruit is in transit, or while in the hands of the commission merchants awaiting a purchaser, or while being re-shipped to other points. The supply in this vicinity appears to be declining instead of increasing with the growing demand, a fact which opens a wider and more inviting field to Southern shippers.

PLUMS

Are really an important crop to Southern shippers. The plum business has been largely experimented so far, and the efforts of cultivators have been confined mainly to the Wild Goose and Chickasaw, the latter a well-known variety which grows in a wild state everywhere South of this latitude. Both make excellent preserves, and are purchased largely for this purpose, especially by the preserving establishments here. A good many, however, hold off before purchasing until the crop from this County (which is generally very large of the Chickasaw) comes in and then buy all required, the prices being at such times the lowest of the season. The Wild Goose is a large showy fruit that finds favor with all buyers. Excellent for either table purposes, retail dealers, or for preserving purposes. It is purchased for a variety of purposes, and the prices are generally remunerative, and up to four years ago averaged very high; but a more general cultivation of it led to a greater abundance and lower prices the following two years. Last year, however, the crop was the lightest for several years, especially of the Wild Goose variety. The Wild Goose seems to succeed everwhere south of St. Louis. Last year Tennessee, Arkansas and Northern Texas shipped some that attracted a great deal of attention on account of their size and fine rich color.

The first receipts are usually from Texas from the middle of May to 1st of June. Mississippi and Arkansas soon follow. Last year's first shipment was on the 15th of May, Texas

Chickasaw, and sold at $1.00 a box. Receipts light for ten days following, quotations being 50c. to 75c. a box and about $2.00 a case 24-quarts. Wild Goose variety did not appear until towards the first of June, when it sold for $1.00 a box, and the Chickasaw suddenly declined to 40c. to 50c. a box. June 15th to 20th Chickasaw 40c. to 50c. and Wild Goose 70c to 90c a box. July 1st Chickasaw 30c. to 40c. a box and Wild Goose 50c to 65c. Home-grown appear about this time and there is little profit to outside shippers after this. However, Southern growers are usually done shipping by this time, if not before, and Southern Illinois follows next, and it is unnecessary to follow the course of the market further. Six-gallon cases, or strawberry packages, are the best to use, though the regular one-third bushel box brings the fruit here very successfully if gathered just at the proper time—and none too ripe put in. A few soft, or too ripe, soon make a bad looking mess of the whole contents of the box. When shipping by freight, gather when full-grown, before coloring sets in, but if by express let them remain on the trees a little longer. A good many arrive too ripe. Many are wanted for re-shipment, hence the importance of shipping before coloring up. They ripen rapidly while in transit, and full ripe fruit should not be shipped, as a few of them sometimes spoil the sale of the package.

APPLES.

Were we not writing for Southern growers and shippers we should open our subject with what Downing calls "the world-renowned fruit of temperate climates," the apple. Apples, both dried, green and evaporated, are shipped as regularly to Europe, as our surplus wheat and other products, and the American apple is steadily growing in favor and popularity in the principal foreign markets. The Southern shippers are interested most in the earliest varieties, such as the Red June, Early Harvest and Red Astrachan, which appear early enough to find our market almost bare. The later varieties, too, should be grown, at least to some extent for local and family wants, if nothing more. Being entirely hardy, producing a crop every year, and thriving with very little care, they can be successfully grown by the most inexperienced. The northern portion of Arkansas is a magnificent fruit region, especially Benton and Washington Counties, which are as far south as we know of profitable orchards, though we are informed several fine apple orchards can be found further south in the State, especially adjacent to the Little Rock and Fort Smith Railroad. West Tennessee, raises considerable apples for the early market. Kentucky and Southern Illinois are largely engaged in the business and contribute liberally to our market. The Red June is the most profitable early variety coming in. The Early Harvest is earlier, but is lacking in color, so desirable an advantage. Apples can be had any month in the year, the old stock appearing regularly until the new crop comes in, and both can be found at the retailer's stand for a month afterwards.

First receipts last season were on the 26th of May, one-third bushel boxes from Arkansas. They sold at 75c. per box—Early Harvest. First week in June, 40c. to 60c. per box. About the middle of June, Early Harvest, 40c to 50.; Red June, 50c. to 75c.; Red Astrachan, 50c. to 60c.

By the first of July, market is full of new apples—boxes are entirely neglected, and selling at 15c. to 25c.—buyers seeking only good barreled stock for shipping purposes, which are quoted at $2.00 to $2.50. Good Red June higher, however, the boxes selling at 40c. to 50c. and the Astrachan next highest. Early Harvest, most abundant and cheap—mostly from local growers—with Southern Illinois next. The season may be said to be at its height for early varieties. A review of the season later would elicit little new or interesting to Southern shippers.

We will say to Western growers that whenever Western New York has a crop it will not pay to store away for winter or spring sales, but when her off-year comes, there is some money generally in putting away good stock for late markets. The New York yield is enormous when it comes—usually flooding every market in the country, besides shipping considerable to Europe.

Earliest shipments, when the market is comparatively bare and prices high, use one-third, half, or bu. boxes, and later, barrels.

Now, in regard to packing: Do not fail to examine all your barrels carefully before packing. Tighten all the hoops, using the shortest nail possible on the bulge. Use liners always at top and

FOR NORTHERN MARKETS.

bottom; do not try and be saving by using inferior barrels. The best are always the cheapest in the end. Apples should always be hand-picked, throwing out all bruised stock and windfalls. These qualities only depreciate the value of your mark and compel you to pay charges on something that has no value in any market.

The facing is a nice part of the trade. Use the brighest and best shaped for this purpose, placing them two layers deep with stems downward. Use those only of an average size, representing the same as balance of contents, so that the buyers are not misled. Do not in any instance, make a facing that will lead the purchaser to suppose that an extra large quality is being sold him, when in reality he will have an inferior lot of stock outside of the few fancy facers. This kind of business is an injury to the house handling your goods, and your brand wll be condemned by the trade. After through facing, finish placing in the balance, by handling carefully in a small way, so as to avoid bruising. Shake the barrel often, so as to have them settle in solid. If intended for immediate shipment, head them up with a Screw Press, always neatly—have an experienced workman for this purpose—so that the barrel after being well packed, nicely stenciled, naming the variety, and well headed, shows up in good, neat shape for the market.

Invariably *mark the faced end* or the one you want us to open for the purchaser. If you mark the wrong end it will lead to confusion and render it necessary to open *every barrel* to ascertain which is rightly and which is wrongly marked. The name of the firm you are shipping to, and the name or variety of apple, should be put on the *head* in every case, with stencil, if possible, but in its absence, with brush or lead pencil. Never leave your commission man or his customers in doubt as to what kind of apple the barrel contains, or which is the top or bottom. Shippers lose a good deal of money by failing to comply with these requirements.

FIGS.

Several years ago we used to get some small shipments of figs from Mississippi. None were received the past two years, and there is really little encouragement to offer shippers. Former receipts, however, sold at $1.00 to $1.25 per box (one-third bu.), but sales are limited, and we can urge only small shipments to this market. Quart boxes and strawberry crates are the most appropriate packages for shipping them.

WATERMELONS.

All the Southern States can grow the melon in the greatest abundance, and as very little skill, knowledge or experience is required, the business is not always profitable in shipping to distant markets. The melons are so bulky, so large and heavy, that the cost of transportation becomes at once the first matter to consider in connection with their cultivation. Your location is

FOR NORTHERN MARKETS. 47

also an important matter. You must be near a railroad station. As melons will not bear express charges, you must avail yourself of the fastest freight accessible.

The first receipts are generally from Florida, Georgia or Texas and appear anywhere from the 10th June to 1st July. Texas usually raises a large crop, but rarely appears early enough in this market to secure early or high prices. Last season first car was from Georgia, getting in 10th June, and selling at $30.00 per 100. Up to the 20th receipts were light and prices $20.00 to $30.00 per 100. About the 1st July receipts more liberal and prices $12.00 to $20.00 per 100. The Kolb Gem variety embraces such a large portion of the receipts that it is quoted separately every day. It is a prime favorite with buyers, and sold higher than anything else offered. July 6th to 10th Georgia melons are quoted $11.00 to $18.00 per 100, fancy large Florida stock $18.00 to $20.00 per 100. July 13th to 15th we find the market nearly bare, and Kolb Gem quoted at $30.00 to $32.00 per 100. Cars of Georgia selling on track for $240.00, and a car from Southeast Missouri—a famous melon growing region, shipped the first car of the season on the 14th July—partly green, small, for $140.00, on track. On the 19th and 20th of July the principal receipts are from Southeast Missouri, and selling from $100.00 to $120.00 per car. At this time Texas is shipping—some of them apparently having just commenced, and they sell for same price as Missouri stock—some lower—or for freight on the average.

We can't understand why Texas, which is fully three to four weeks earlier than this latitude, should be so late getting into this market. Shippers must understand if they can't get here until this State commences to ship, they had better avoid this market. If they desire to get out of the melon business what it is capable of yielding, they must take some chances in getting their plants out early, and taking good care of them afterwards. There is no good excuse to offer for getting in so late.

After this State gets thoroughly started, there is no room for more distant shippers. August the first, melons are quoted at $40.00 to $80.00 a car on the track. Later you are not interested in the course of the market.

In packing, hay or straw, or similar packing material, should be spread over the bottom of the car. Thoroughly ventilated or cattle cars, should be used, the sides if open, protected against pilfering, by nailing planks on the inside or openings. Many melons are too ripe when shipped, and it is equally true many ship too soon, when the melons are far from full grown.

CANTALOUPES.

The supply within reach of this market the past two years was not large. They usually appear a week after the advent of melons, but the first receipts are eagerly sought and usually bring $2.00 to $2.50 per dozen for a short time; after that, the price steadily declines, until car-load shipments are made, when they sell by the hundred, dropping as low as 3c. to 5c. each, sometimes. Barrels (chipped or ventilated), boxes, crates, etc., will do for early shipments. The regular melon crate made for this purpose, holding from one to two dozen, according to size of melons, is the most appropriate. Put in no overripe, specked, bruised, damaged or faulty melons. Neither must they be picked too green or half grown.

The first receipts last season appeared from Mississippi, on the 17th of June, and sold at $2.50 per crate. Arkansas commenced on the 20th, selling at $2.00 per crate, at which the market remained for nearly two weeks. First week in July, quotations are $1.50 per crate, and most of the receipts from Arkansas. July 15th, Missouri growth is among the receipts, and prices are $1.00 to $1.50 per crate. After this date the prices steadily decline as home-grown receipts increase, until the 1st of August. We find them quoted at $3.00 to $6.00 per 100. White Japan, Bay View and Nutmeg, are the most favored varieties, in the order named.

Most of the Mississippi receipts came through in good order, by through, or fast freight trains. If picked at the proper time, just when full grown, will stand two or three days easily.

Miscellaneous Matters.

ABOUT IRRESPONSIBLE HOUSES.

There is no more appropriate place to devote a few lines to such a subject, and we cannot ignore the opportunity to offer a few words of explanation and caution under this head. Shippers are so situated that it is difficult for them to ascertain at short notice the standing or responsibility of certain firms, and the question rarely occurs to them, until they are ready to ship, or appealed to for business by some new firm they know nothing about. They occasionally receive a letter soliciting shipments, that is so alluring and tempting in character and make-up that some cannot resist it, and take the chances. You are told of the prices they can secure or guarantee you, the advantages they have over all other houses, and all the tricks to catch the unsophisticated are resorted to and very often successfully.

Now, fruit shippers must understand that every mercantile business has a certain number of adventurers, men who have everything to make and little to lose, and who some day after having secured the confidence, patronage, and funds of the confiding and unsuspecting class, disappear as suddenly and as unexpectedly as they appeared. The fruit commission business, we regret to say, is not free from this class any more than any other calling, and never will be, and hence the importance of intrusting your business to well-known, experienced and long established firms, for if you cannot do well with such houses, what show will you have with a different class? In every large city a few of such firms are apt to come to the front, making their advent with the first arrival of fruits, and too often, when the fruit season is over and no more to be made, they go down with the leaves in the fall. You are surprised how they got your address. This is simple enough, as they can be secured from the packages in front of the commission houses, or at the express offices every day. We know of a great many fruit shippers who will appreciate the force and wisdom of the foregoing remarks, but we are writing for the benefit of the less experienced, that they may avoid the expensive experiments of others and profit by their experience. These remarks will apply with equal force to all markets as well as St. Louis, and this little volume represents so many of them that our friends will patronize, that we urge the greatest caution, since few shippers are in circumstances to stand the losses frequently arising from shipping indiscriminately.

ABOUT COMMISSIONS.

The inexperienced shipper often objects to ten per cent. commission, the universal charge in all the principal cities, by firms making a specialty of such products. There is, perhaps, no business requiring so much stationery, writing, stamps, stencils, drumming expenses and preparatory work as ours, and to do justice to these very perishable products, you can really attend to nothing else while they are coming in. As a matter of fact, no merchant gets ten per cent. for selling the goods, for fully five is consumed in the cost of soliciting, whether by local agent or traveling man, coupled with the cost of stencils, reports, etc. It would be much easier and more profitable to sell other goods over which you need not be so exercised at five per cent. The most forcible argument, in favor of the justice and propriety of these rates of commissions, is the action of most of the oldest, largest and most experienced shippers, who will not ship to any firm who charges less, and at the end of the season the wisdom of their actions will be apparent.

DIVIDING SHIPMENTS.

Fruit growers frequently divide up their shipments too much. We have in our travels often seen shippers mark half a dozen packages to three or four houses. This is all wrong, and rarely pays as well as if shipped to only one or two houses. The same labor and amount of book-keeping is required to record and report these little shipments, as large ones, increasing the opportunities to make mistakes at each end of the line, giving as it does double work to express agents at both ends of the line, frequently increasing the express charges, and requiring so many more reports from here, stationery, postage stamps, price currents, etc. All of which go to show the practice is ill-advised—doubling the labor to all concerned without any benefit in return.

The most difficult man to satisfy is the new shipper. He

expects too much generally. He has an idea that you await, with some anxiety the arrival of his shipment, also a number of your customers. He will expect a long letter giving the full details of its condition, etc.; what it brought; and if the returns fail to come up to those of any of his neighbors, you have made an enemy in most cases, and he is ripe for a change and an easy prey to the first drummer that comes along. An explanation, if you have time to make such, rarely satisfies him. The commission house soliciting the new shipper will find a number hard to please; we know this from long experience. And as shippers and receivers are looking for all the information they can acquire, we are reserving nothing through selfish or other motives from either party in this work.

SOME POPULAR ERRORS.

The belief prevails widely that fruit commission merchants are rich, have an easy way of making money, and steadily adding to their wealth at the expense of the shipper; that they are in a safe business and should never fail. This is a rosy picture and we wish it was only half true. Then, reader, let me tell you, they are not rich, and it can be said with the greatest truth they are not in the right sort of business to become rich, or even in easy circumstances! There is not one of them, in this city at least, practicing anything but economical habits either in living or conducting business, and they are remaining in the trade in the hope that the future will prove more profitable than the past. In the hot contest for existence in commercial life now, the man who succeeds in meeting his current expenses and bills promptly is exceedingly fortunate, and the firms that have a little balance on the right side at the end of the year are few—exceptional in fact. There are some commission houses who have made some money the past ten years and who still hold it, but none of it accumulated from handling fruits and vegetables on commission. It has been the result of lucky ventures or speculation in other directions. No firm in this line of business can live solely on consignments of fruit, etc. All must handle other products at least six months in the year, for the fruit season pays expenses only while it lasts—no longer.

The commission man suffers from bad debts, as all others do, for two-thirds of his sales are charged up, and no matter how many bills he loses, the shippers must be paid for the goods. If all were sold for cash the receiver would have to take a great deal less for his goods, and the shipper would as a result get smaller returns. The seller, in his anxiety to please the shipper, beats his rivals and build up his business, often takes chances in this way he afterwards regrets. Many years ago when the force of competition was not felt so keenly, there was more to be said in favor of the business, and no less inviting field exists at present for a man desiring to go into business. Many have tried it here and elsewhere of late years to their sorrow. They found trying to do a paying business competing with old established and long experienced firms next to impossible; that not only considerable money but also a wide experience was essential to success in the undertaking, and that it required several years to even secure a paying patronage. We have in our long experience seen so many young men, generally off-shoots of old firms, start out full of hope and soon after find oblivion, that we are competent to write at length on the subject, and hint advisedly in the foregoing remarks.

THE DRUMMING QUESTION,

Is one of the most provoking to the receivers as well as shippers. It imposes on the commission men a heavy tax they have in vain sought to avoid. At fruit-growers' conventions and meetings, the subject comes up for discussion occasionally, and resolutions adopted, setting forth, that houses hiring drummers or local solicitors, will not be patronized, etc. It is equally true, that later or before the first case of strawberries is ready, it takes only the eloquence of the average drummer to secure it for his house, and the foregoing resolutions, subscribed to by the shipper, wholly ignored. When the season is fairly under way the best solicitor or talker, no matter how poor or irresponsible his house may be, usually succeeds in getting the most, at least for a while, until the returns begin to come in, when he can be found equally industrious at other points, and thus while working up a fine business keeps out of the reach of disappointed shippers. .

The result is, every firm, no matter how old, reliable, or responsible, or how good figures it can secure, will get left unless it has a man on the grounds to fight for his share. Thus, in self-defence, he is forced to hire a solicitor and place him where the shipper says (at the meetings) he is not wanted. So the shipper after all, creates the evil he complains of, and sustains it.

RECEIVERS UNJUSTLY BLAMED.

As a sample of how commission men can be unjustly censured, we will relate a little experience of our own which occurred last fall: One of our Missouri apple shippers, whom we esteem very much for his liberal patronage, made a shipment which we reported same day received—as in bad order, slack barrels, specks, faulty fruit and bad packing generally. Next day, on receipt of the letter, he replied, we were certainly mistaken, that it was not his fruit we were writing about, etc. We telegraphed him to come down on first train, and if we were wrong would pay the expenses of the trip, and if he was wrong, he could do so—an offer he accepted. We showed him his fruit, which he admitted was his, opened some barrels not yet touched, and found them about same as those complained of. He expressed a great deal of surprise at its condition and how it depreciated in value in such a short time. He saw then very forcibly the result of rough handling of fruit which should be carefully hand-picked, and the poor economy in hiring cheap, green hands for the picking and packing of his apples. He admitted the fault lay chiefly with the help, who did not follow his instructions in packing. His trip paid him and ourselves too.

YOU CHARGED ME TOO MUCH!

Every receiver has this charge frequently thrown at him by more or less angry shippers, whenever the railroad company or express company makes a mistake or overcharge. Now, we have nothing more to do with the making of these charges than the man in the moon. We have to pay whatever the express company or railroad company demands, and if there is anything, file our claim, which is investigated and straightened out afterwards.

SLOW RETURNS

Arise from various causes. Packages come in occasionally with no mark visible to indicate who the shipper is—the tag or card torn off. If no advice by mail, the receiver has to wait until he hears from the shipper. The importance of advice by mail is manifest here, or better still, placing in package a slip or card showing your address and contents, or call for a stencil which we will mail free for marking.

When goods come by freight, the most provoking delay is that caused by the railroad companies centering on the east side of the river. All lots smaller than car loads are turned over to transfer company on arrival. They deliver the goods, but the freight bill never shows up for about three days. The transfer company is never prompt collecting these bills, as they should be, to enable us to report promptly. The railroad companies on arrival of goods turn them over with freight bill to the transfer company, and the transfer charges are added to the regular railroad bills—an extra charge that is not understood nor figured on, by many shippers—and letters of explanation in regard to these charges are often called for.

HINTS TO SHIPPERS.

A number of shippers, the new ones especially, when they receive a stencil, regard the number on it as the street number of the firm sending it out. This number really represents the shipper's address—being placed in the books opposite his name as soon as sent out. His address on the package in addition to stencil number is therefore superfluous.

The stenciling should be on the cover of the package, serving as it does, to keep the right side up. Such packages as strawberry cases should also be branded on both ends.

If you have no stencil, a lead pencil can be used to write the firm's address, and your own should follow, writing the word *from* between them.

A shipper frequently borrows his neighbor's stencil, and

uses it without notifying his commission house. You can see how this will complicate matters. Your neighbor will get the returns, and if he refuses to settle with you the commission house must pay twice or incur your everlasting displeasure.

If there are any empty boxes in a crate, always make lead pencil note of same on cover; and if two or three varieties are in same package, as is sometimes the case, indicate it in the same way.

In the midst of the fruit season every commission house is driven to death, and has no time to either write or ask for explanations. If you do not hear from your shipment promptly, you may consider something is wrong; so send in a few lines, asking and giving explanation in connection therewith.

When shipping by freight always notify consignee by sending receipts or otherwise.

Never use large or irregular nails for fruit boxes or crates; such spoil the appearance of the package and injure the sale.

A common error by shippers is that of waiting too long before ordering their fruit boxes. They are often detained on the way, and frequently the box factory is crowded with orders and you must wait, and your fruit is spoiling in the meantime.

The cost of stencils is quite an item to commission merchants—those with your address 15c. to 20c. each, the numbered ones 5c. to 10c. each; so do not destroy or lose them. Some shippers call for new ones every year, as if they cost nothing.

Whenever practical, fruit should be shipped at night or in the evening, getting in this way the benefit of the cool atmosphere while in transit. Getting to our market in the morning, early as possible, is also an important consideration.

Saturday is always the poorest day in the week to sell to advantage, as few shipments are made on that day. Friday is the best day in the week—the shipments on that day being heaviest.

Avoid as far as possible getting goods into market on Sunday morning. They will keep much better in the country than in the city. Monday morning the market is usually bare, and Sunday night shipments strike a good market generally.

TRANSPORTATION CHARGES.

This is a very important subject to the fruit grower, especially if he is an extensive shipper. The cost of transportation has materially checked the cultivation of fruits and vegetables in many sections where all other conditions were favorable to the enterprise. Express rates in some instances amount to a prohibition, where there are no competing lines or companies. Take certain shipping points in Southwest Missouri for instance, where it costs much more to market fruits than it costs a large number of the Arkansas shippers situated on the St. Louis, Iron Mountain & Southern Railroad, much further from our market. The result of a lively competition is nowhere more manifest than on a railroad, in the matter of rates.

SPECIAL RATES can always be had on perishable goods from the express companies. New shippers, located at new points, where special rates are not established, should avail themselves of this advantage, and ascertain the lowest rates they can secure before they commence shipping. There is a marked difference between special and regular rates. The Florida shippers, for instance, have this season from $3.00 to $3.75 per 100 special rates, while the regular is $5.00 to $6.00 to this city.

DRIED AND EVAPORATED FRUITS

We handle regularly, and our shippers can rely on getting at all times the best figures the market affords. Barrels are the most appropriate packages, although sacks and boxes can also be used. This latitude being much earlier that the more Eastern States, where most of the dried and evaporated fruit comes from, as well as the green fruit, shippers will find it to their advantage to ship as soon as ready for market, and not wait until competition springs from points further East. New York State, which furnishes *three-fourths* of the evaporated fruit of the country, finds St. Louis a profitable market; and as your climate places you several weeks earlier in the market, you should profit by this pportunity to sell while the markets are comparatively bare.

TO SHIPPERS OF DRIED FRUITS.

Apples should be carefully peeled and cored, then sliced or quartered, placed upon frames and dried in a gentle heat. Gnarly or wormy apples should be thrown aside, or such places carefully cut out. Peaches may be dried either peeled or unpeeled. They sell best if cut in halves. Cherries must be pitted, and to bring good prices they must be very dry, entirely unmixed with sugar. Black raspberries and blackberries are dried whole, and care must be taken that they are not crushed and broken. Apples and peaches, to bring best prices, must be bright and light-colored; to secure this, they must be dried in a dry air. The atmosphere is often so charged with moisture, even in sunshine, that it absorbs more moisture very slowly. Such an atmosphere is very unfavorable to the drying of fruit, the juice evaporating so slowly that it decays and darkens the color. Those who cannot construct drying houses should prepare and dry their fruit upon days when the air is very dry only, out-of-doors, or else in-doors in a gentle fire heat and current of air. Apples on strings are objectionable. If dried on strings, these should be removed before the apples are packed.

EVAPORATING.

Bleaching is done by exposing the fruit in a wooden box or special machine, to sulphur fumes. The sooner the bleaching is done after the apples are cut the better. Caution is necessary not to overbleach the fruit and cause it to both taste and smell of sulphur. In different establishments the heat of the evaporator varies from 95 degs. to 175 degs. Fahrenheit. The fruit must remain in from two to five hours, according to the heat of the air in the evaporator. One bushel of apples is estimated to make from five to seven pounds of dried fruit.

BREAKING DOWN THE MARKET.

Remember the market is never broken down by *good* fruit. It is the great quantity of *poor* fruit that oppresses the market and forces down prices. We are as interested in sustaining the market and prices as you are, because when prices are away down we get nothing for our labor, and hence we urge more good stock and less poor and indifferent stuff. How much more profitable and satisfactory to get $20.00 net proceeds for ten packages fruit, than to get only the same sum for twenty packages. You are out the packages, labor, etc., and the express companies only have profited by the enterprise. This is a fair illustration of the case, no matter what you ship.

CANNING FACTORIES' ADVANTAGES.

A canning factory is a more important adjunct in connection with truck-farming than can be seen at first glance. In any town where the business is conducted to any extent, it becomes a paying institution in a two-fold sense. Say there are twenty merchants who will contribute $150 each. This $3,000 will furnish all the machinery, or plant necessary to do a large business. As the canning is conducted during the summer any cheap structure will do for a building. One hundred hands, women and girls, can be profitably employed for several months, in such a concern, and it is safe to say the bulk of the earnings of this one hundred go right into the hands, or business, of the twenty merchants who invested in the project, and even if they lost the principal, the increased income, business and population resulting from the erection of the cannery would be largely beneficial to all concerned.

It should be added that the new man casting about for a location to embark in the business cannot but regard this factory as a promising and encouraging auxiliary, since he is assured about first cost for that portion of his products it may not pay to ship. A good deal more might in justice be said in favor of the project.

Tomatoes, corn and peas are the staples in the vegetable line, and strawberries, peaches, plums, etc., in the fruit line.

Next in importance would be a box factory, which could be secured at a similar cost, and most of what has been said in favor of the canning establishment will apply here.

Crystal Springs, Mississippi, concluded to erect both such establishments this spring, and the result has been a regular boom, and quite an increase to the population of the town.

COLD STORAGE

Is an important subject to many fruit growers. The man who has access to such an establishment can occasionally use it to good advantage in tiding over crowded markets and low prices. Apples, pears, grapes, etc., are fruits that can often be profitably stored. In putting away fine specimens, or lots of any kind fruits for fairs or exhibition purposes, the right kind of cold storage becomes an important help. St. Louis has been for years sadly in need of a first-class establishment of this character, but now has one, equipped with all the modern improvements and machinery essential to success, and any temperature from zero up can be furnished patrons at any time. See the firm's card in this work for further particulars, rates, etc.

REFRIGERATOR CARS.

For some years we had an idea that the refrigerator car, properly managed, could carry fruit as perishable as strawberries and peaches to markets a thousand miles away. Impressed with this belief we concluded to try the experiment three years ago. When our home-grown berries were coming in freely, we purchased one day in the market 150 stands (four drawers, or two bushels to each stand), paying for the same $5.00 each, or $750 for the load; they were all good *Wilsons*. That same evening we saw our car started over the Chicago, Burlington and Quincy Railroad, bound for Denver. It was the first refrigerator loaded with berries that ever left this city for any destination. We were compelled to prepay the freight on same, of $250. So to test our judgment in the project, we had to put $1,000 in jeopardy. The officials assured us they would take special care of this car along the route, and that the car would be regularly iced at the various points where such arrangement existed. However, we had not entire confidence in the promises made us, so sent our man along with the car to see that the ice did not run out between the icing stations, a most fortunate provision, as the sequel will show. The second night out, although up in Iowa, was exceedingly hot, and our man discovered that the ice was about gone, and the next stopping place far away. In the meantime the rising temperature in the car threatened serious results, so, with the assistance of the train conductor, telegrams were sent ahead to have ice at the first stopping place. Being a through fast freight, few stops were made or allowed. The ice was secured as requested, and the car re-iced before any injury was inflicted on the fruit. The contents were duly received and delivered at store of consignee in fair order. The time involved between this city and Denver is usually sixty to seventy-two hours.

West Tennessee and Southern Illinois have shipped to Eastern markets in the same way successfully we think. At each place, however, a cooling house has been built, where the fruit is thoroughly cooled off before loaded into a refrigerator car, and this is a very important and valuable auxiliary to those desiring to ship to distant markets.

We had, three years ago, from Texas, several cars of peaches (refrigerators). The fruit, though out three to four days and nights, reached us in good order, and experienced, not only while in transit, but after arrival here, unusually hot weather, the bulk of the fruits selling at 70c to 75c per box.

Contrary to the general belief the fruit did not melt down, or become discolored, after being exposed to the warm atmosphere

We had some of the fruit on hand for one day additional, and still it looked well; the whole lot selling to as good advantage, or nearly so, as if it had come by express. The varieties, however, were judiciously selected and gathered apparently at the proper time to stand three or four days' ride. They were partly Chinese clings, good size and color, and quite firm. The other varieties were composed of good, yellow, firm fruit, and none were really in bad order. The good judgment used in selecting and packing the fruit brought about in a measure the success reached.

The great danger in the use of refrigerator cars is the running out of ice while in transit. The amount of ice required, depends largely on the weather. A quantity that would be ample for one week, may prove wholly inadequate the next week or previous week. While the usual supply provided by railroad companies may meet the average requirements, it will not do always as in the case of car we shipped to Denver, which might have been ruined had not our man been along to watch and protect it.

VEGETABLES.

The fine onions, potatoes and tomatoes that appear in the leading markets in January and February, the first receipts of the season, come from Bermuda, a small coral island in the ocean, or rather a series of islands strung together, embracing an area fifteen to sixteen miles long, by a mile to two miles wide, form the Bermuda island, which is situated several hundred miles east of Florida. The whole island is given to the cultivation of these crops, the people, some sixteen thousand, relying on the outcome for a living. A feature of the business there is the law which compels every producer to put his name on every box or package, so that any crookedness or deception in packing can be traced to the proper party. The purchaser, therefore, never has to open more than one package to examine the goods. Onion growing is remunerative there, realizing about $400 per acre. They are put up in substantial boxes, holding fifty pounds each. Tomato growing there is equally profitable, securing the grower $400 to $500 per acre. They do not grow as large as those furnished from the States—the majority of them being below the average size. When about full grown and beginning to color up, they are gathered, wrapped in coarse brown paper, and packed in seven-quart boxes (generally considered pecks).

The packages used there are thoroughly ventilated—the sides being composed of slats. The products of the island going into the States in such cold weather, hold up in good order four to eight weeks, and command high prices, having everywhere during the greater part of the time, no rivalry. Florida, however, is only a few weeks behind with some of her products. The St. Louis markets, however, received but very little of such goods heretofore, as they found a cheap water route to the large populous cities of the East, where a majority of them are consumed. New Orleans and vicinity are next heard from, and soon afterwards Southern Alabama, Mississippi and Texas, or sections of those States contiguous to the coast, commence shipping here. Florida products, however, made considerable of a display in our markets last spring, and will increase her shipments here in the future, encouraged by better railroad facilities and lower express and railroad rates.

MOST PROFITABLE VEGETABLES.

The staple or best paying vegetables for our market embrace cabbage, Irish potatoes, peas, beans, cucumbers and tomatoes. Most of the other vegetables consumed here are furnished very early by the hundreds of gardeners in this vicinty—almost as soon as wanted, and we could not recommend the cultivation of anything else except in a very limited way. They are referred to further on in this work.

CABBAGE

Can be found in our market steadily throughout the year, the old and new crop appearing side by side in February, March and April, and the price largely governed by the quantity of old cabbage on the market. The first receipts last year were from New Orleans, and sold at $5.50 per crate. The market was nearly bare at the time, and for several weeks previous to this, there was no cabbage here to speak of except the California stock, which was coming by the car-load and selling at 3c. to 4c. per pound. The prices slowly declined until the 12th to 15th of April, the quotations are $3.50 to $4.25 per crate, and receipts mainly from New Orleans and Mobile. From the 20th to the 25th, prices of Louisiana receipts are $3.00 to $3.25, and Mobile, $3.50 to $4.00 per crate. The first week in May disclose no change—the figures being $3.00 to $4.00 per crate, according to quality, etc. May 15th to 18th, Louisiana, $2.25 to $3.25 per crate, and Mobile, $3.50 to $4.00 per crate; 20th to 28th, price unchanged—$2.25 to $4.00 per crate. First June, home-grown, has appeared, and sells by the crate in shipping order at $4.00, with prices of Southern unchanged, owing to light receipts. June 6th to 10th, home-grown, $3.25 per crate, and shipments, $2.00 to $3.00 per crate. It is useless to pursue the course of the market later, since Southern shippers are no longer interested, nor would it be profitable after the 1st of June. The weather at this time is very hot, and considerable of the receipts in bad order, through decay, etc., especially where good judgment and proper precaution was not exercised in packing.

As the weather grows warmer, greater care must be exercised. It should be nicely trimmed, and though a few more heads are necessary to fill the crate, the work will pay well. It must be tightly packed, for the natural shrinkage, especially in such weather, is very great, and no inferior, wilted, or damaged heads should be packed.

CUCUMBERS.

The first receipts appeared about the middle of February, and sold at $2.50 per dozen. Of course, sales are very limited at these figures. By the 25th, prices are down to $1.50, and on the 1st of March, $1.25 is the market, and receipts mainly from Florida and New Orleans. March 10th to 15th, 75c. to $1.00 per dozen, and a few days later, receipts are noted from Texas and Mobile. On the 1st of April, a portion of the receipts, so poor as to grade culls, are quoted at 50c. per dozen, and good to choice stock, at 75c. to 90c. per dozen. 15th to 20th, owing to continued light receipts, prices choice are unchanged, but the majority of the receipts are now inferior and selling at 30c. to 50c. per dozen, with choice 90c. The first week in May, find the extreme range at 40c. to 75c. per dozen. May 12th to 15th, 30c. to 5Cc. per dozen, the receipts from half-dozen Southern States. The home-grown appear about this time, and Southern growers are no longer interested.

Never ship a yellow, wilted, stale, overripe, stunted, or half-grown, or over-grown cucumber, unless you want to spoil the sale of all others. Ventilated boxes or barrels can be used in shipping. The *Improved White Spine* is the most extensively grown variety for commercial or shipping purposes. If they are cut instead of being pulled off the vine, both the cucumber and vine will be benefited thereby.

GREEN PEAS

Usually appear here the latter part of February or 1st of March. Last year first receipts were on the 3rd of March, 3-peck boxes, from New Orleans, and sold at $3.00 per box. March 6th to 10th, $2.25 to $2.50 per box; 15th to 20th, Mobile receipts, $2.50 to $2.60 per bushel box, and New Orleans, 3-peck boxes, $1.50 to $1.75. First of April. Mississippi and Alabama receipts, $2.50 per bushel box, and one-third bushel boxes, 90c. each. Weather at this time (frequent showers, etc.,) most unfavorable, and a portion of the receipts in very bad order, those not dry and cool as they should be when packed. 5th to the 10th, majority receipts from Mississippi, and the quotations, Mississippi $2.25 to $2.50 for bushel, and one-thirds 75c. to 80c., showing the smaller boxes most salable—being generally in better order—especially during unfavorable weather or delay in

FOR NORTHERN MARKETS.

transit. New Orleans $1.50 to $2.00 per 3-peck boxes, and Alabama receipts, bushels, $2.00. About the 15th Arkansas is shipping and getting 90c. per one-third bushel box, and prices from the others show but very little decline. April 20th, Mississippi receipts largely in bad order, the result of a prolonged drouth. Large boxes $1.25 to $1.75 and one-thirds 40c. to 65c., according to condition, Arkansas 90c. to $1.00 for one-third bushel boxes. On the 25th to 1st May we find no change in figures. May 3rd to 8th, Arkansas 75c., Tennessee 75c., Southern Illinois and Southeast Missouri 85c. The more Southern shipments in bad order and receipts falling off rapidly. Home-grown appear soon after this time, and the course of the market will not interest shippers much longer.

Green peas are generally a profitable crop, for Southern growers, when properly handled. The importance of getting them here green and fresh, and in neat, ventilated boxes, one-third and one-half bushel boxes (the former preferred), cannot be too strongly urged. A number of these packages are re-shipped; such bringing the best prices, but only the freshest and nicest looking stock, properly packed, go to this trade. Peas that are overripe, discolored or wilted, as some of the receipts appear, are almost unsalable in any market. Another great mistake is that of picking too soon, before half grown or half full. The past year a number of the boxes were poorly made, the openings frequently permitting the peas to drop out freely every time the box was moved. The peach box being generally used for this purpose. sufficient care was not exercised; though the same material, properly split, will make a good pea box. They heat readily in large packages, especially in barrels, sacks, or tight packages, even when shipped by express. They should not be out over two days, or three days at most, though they cannot be regarded very green or fresh if on the way longer than twenty-four hours, in the warm weather usually prevailing at that time.

In packing shake down thoroughly, and a little pressing down in nailing on the side piece or cover of the box won't hurt them. Have them as cool and dry as possible before packing, to avoid heating.

STRING BEANS.

The first receipts arrived on the 25th March, three-peck boxes, which sold at $3.50 per box. Receipts continued light for a week or more, New Orleans and Florida shipping. On the 1st April $2.25 to $2.75 per box for freight receipts, and express $2.50 to $3.00, mainly New Orleans 3-peck boxes. April 10th prices are about the same and bulk of receipts from Louisiana. On the 20th receipts quite liberal and prices $1.50 to $1.75 for 3-pecks. On the 25th prices are unchanged, but on the 1st May better stock and higher prices—Alabama $2.00 to $2.50 and Louisiana receipts $1.50 to $2.00. Five days later 25c. per box lower all round. The wax variety has appeared and sells 25c. per box higher than the green. May 15th report reads, New Orleans flat 75c. to $1.00 per box and wax $1.50, Alabama round $1.50 to $1.60 per box (bushel). May 20th New Orleans 90c. to

FOR NORTHERN MARKETS. 75

$1.00 for green, wax $1.65, Alabama $1.25 to $1.50 per box, wax $1.75 to $1.90. 25th, Louisiana $1.00 per box for green and wax, Mobile $1.25 to $1.50, Arkansas 40c. to 60c. for one-third bu. boxes, and wax variety at 65c. per box. A week later home-grown appear and secures most of the trade.

The round bean sells much better than the flat variety, and the Wax bean generally higher than either, though the market will not consume near as many of the latter. The flat (Early Mohawk) is the earliest and most valuable on this account. The Valentine, or round bean, is tenderer and less stringy, and sells higher. Pack same as the pea—though they do not shrink as much as the pea while in transit.

In packing exclude all the moisture possible and let them be as cool as circumstances will permit. With proper precaution, so many will not reach here mouldy. In unfavorable weather they carry and keep better in one-third or half-bushel boxes—although the majority of the receipts come in three-peck and bushel boxes. A good many of the Arkansas beans come nicely packed in one-third bushel boxes, the beans crossing all the openings, so that none can drop out in this way while in transit.

TOMATOES.

As stated elsewhere, the first receipts are from the Island of Bermuda and come into the leading markets, East and West, during January and February. They come wrapped in brown paper, it seven-quart boxes, and keep in good order for several weeks, the price being usually about a dollar a box. We judge they are gathered when full grown, and then given plenty of time to ripen and color up while riding around the country. The first receipts last year appeared from Bermuda and Florida about the same time, 1st February, and sold for $1.50 peck boxes. The market remained quiet and steady for a week, when they declined to $1.25 per box. 10th to 15th they average about $1.00 per box. March 1st Bermuda and Florida shipping steadily and prices are lower, 50c. to $1.00 per box. March 15th to 20th receipts are light and prices $1.00 to $1.25 per box. April 1st to 25th only Bermuda stock are offered, and scarce—$1.50 per box. First week in May receipts still confined to Bermuda growth and prices $1.00 to $1.25 per box. May 12th to 20th Bermuda 75c to $1.00. It is a very unusual thing to find none of the Southern States shipping tomatoes here late as the 20th of May. On the 23rd first receipts from Mississippi, which sold readily at $2.50 per one-third bushel box. Receipts for several days are mainly from Mississippi, but too green to sell at full or ripe prices—$1.75 to $2.00. 25th, Alabama ships, getting $2.00 per box, and a few days later Texas and Louisiana are shipping, and prices range from $1.50 to $2.00 per box. 1st June, bulk receipts from Mississippi and Alabama and prices $1.50 to $1.75 per box. June 3rd to 6th, large receipts. Florida $2.00 to $2.40 per bushel crate, Mississippi, Texas, Louisiana and Alabama 75c. to $1.00 per one-third bu. for stock not quite ripe—full ripe $1.25. Crystal Springs Mississippi, where vegetable growing is conducted on a very extensive scale, commenced shipping at this time; by freight in car-load lots. The stock was very fine, the freight receipts arriving in splendid order, and sold right along at higher prices than were paid for most of the receipts from other points. They were gathered at just the proper time to stand the two or three days, were properly packed, and were eagerly sought for by the general trade. Shipped nearly a car each day for some time. June 8th, Arkansas commenced, getting $1.00 per box, the highest prices ruling then. June 15th to 20th, Mississippi 50c. to 75c., Arkansas 75c. to 90c., Southern Illinois, Southeast Missouri and Kentucky 85c. and 90c. per box. Home-grown commenced to come freely soon afterwards and shippers will have to gradually pull out. Southern Illinois, however, shipped profitably for a long time afterwards.

The importance of proper handling, packing, etc., is not properly estimated. If ripe tomatoes are going to be gathered, be sure you put them in a separate box; but ripe stock should not be sent forward unless you are only a few hours' ride from market. Even then they are liable to arrive in bad order.

Generally speaking, the proper time to gather and pack is when the tomato is full grown and beginning to color, or partly colored, depending upon the time in transit. The warm weather prevailing at the time will ripen them fast enough. You should not lose sight of the fact that a good many are wanted for re-shipment, and to be fit for this trade, the best we have, must not be fully ripe when they reach us. When shipped by freight they must be gathered still sooner, when full grown, before coloring sets in. Freight is not desirable unless you have some assurance in regard to time. A good many come from the South by freight that are almost worthless on arrival. Last year considerable came entirely too green; that is, were picked and shipped before *full grown* and most of such stock arrived rotten. The regular peach box (one-third bushel) should be used.

The best packing usually appearing in this market, is that from Southern Illinois, where the most experienced growers reside. Their packing is almost perfect. No knotty, stunted, overripe, or otherwise imperfect stock should be put in the box under any circumstances there. The receipts from that section are always sought by the shippers here in consequence. There is a very wide demand for the tomato; all classes being purchasers as soon as the price becomes reasonable. The demand for it is steadily on the increase.

A great deal of money has been made off the tomato, not only in the South, but also North and East. The South is destined, however, to remain the most profitable region to cultivate them for commercial purposes. The improved facilities and lower rates for reaching Northern, or distant markets, continue to afford substantial encouragement. An acre of ground can be made to yield enormously in efficient hands; from one hundred to four hundred bushels, according to circumstances, location, etc. Two hundred is, perhaps, the limit in the South, and four hundred in the North.

As to varieties, will say that the "Acme" should head the list for this market and for most other markets, too, as it is a universal favorite. "Livingston's Perfection" possibly next, and any smooth, round, medium-size variety, might be added.

IRISH POTATOES.

The first receipts, anywhere from March 1st to April 15th. Last spring, first receipts March the 10th, from New Orleans, and sold at $3.00 per box, 3-peck boxes. First receipts this spring from Canary Islands, via Mobile, March 5th, and sold at same price—$3.00. A year ago the new supply found the market supplied abundantly by the old crop, and prices declined rapidly. March 17th to 20th, $1.50 to $1.75 per box, and $5.00 per barrel; on the 25th, 75c. to 90c. per box and $3.00 to $3.50 per barrel. April 1st, scarcer and higher, 75c. to $1.00 per box, and $3.50 to $4.00 per barrel, with receipts mainly from New Orleans. April 10th to 15th, most of the receipts in barrels and selling at $3.25 to $3.50 for good to choice. A good many small and inferior, sunburnt, etc., coming, and selling about, $2.50 per barrel. On the 20th to 25th, the market is $2.50 to $3.00 per barrel.

On the first of May, a better demand exists for new, the old supply being more neglected and prices are higher, $3.00 to $3.65

per barrel, and choice sacked, $1.10 to $1.40 per bushel. Most of the barrels slack. On the 10th to 15th, a steady advance is noted, $3.75 to $4.00 per barrel, and on the 20th, $5.00 per barrel is paid freely for good stock. June 1st, $4.00 to 4.50 per barrel. June 2nd, first home-grown receipts appear. Receipts from all sources continue so light that no decline follows for a week or more, but by the 15th of June, receipts very liberal, and they are quoted at $2.50 to $3.00 per barrel, and by the bushel, $1.00 to $1.10. The Southern shipper is no longer interested, as prices steadily decline as home-grown continue to increase.

In shipping by car-load be sure to secure a well-ventilated car. Occasionally cars come into the depot leaking—the contents nearly worthless. The weather was exceedingly warm, close and damp at such times, and the cars were almost air-tight, and the result was a serious loss to shippers.

For early shipments the packages must be thoroughly ventilated, whether boxes or barrels are used, and should be well filled so as to prevent shaking while in transit. The unripe or those not fully matured are easily bruised, and soon become so discolored as to spoil the sale of all. A number of growers ship a little too soon, and lose money by doing so.

It is very important that potatoes should be barreled as soon as possible after they are dug, as lying in the sun heats them and causes them to rot. Avoid digging immediately after a heavy rain. All potatoes should be barreled when as dry and cool as it is possible to have them. Assort very carefully and ship nothing but the largest, having them as uniform in size as possible, as culls or small ones do not increase the bulk much, but add to the weight and damage the sale, so that they bring no returns and actually depreciate the value of the full size ones with which they are mixed. Use full size, well-ventilated barrels, fill to heaping, and shake down thoroughly.

Early Rose, Early Ohio, Beauty of Hebron, Burbank, and Peerless, are standard varieties here, as they are at most other points.

SWEET POTATOES

Brought better prices last year than for the two or three years preceding, on account of light receipts and the good price prevailing for Irish potatoes. We have every year a heavy correspondence from parties throughout the South, who want to ship here, attracted often by the good prices ruling here, which usually apply to home-grown stock, and which always sell much higher than shipments from the South. The Southern receipts are always more or less discolored through bruises and injuries acquired through packing, rough handling, or shaking up received while in transit, and not infrequently dry rot visible to some extent. The large size and somewhat coarse texture, coupled with lack of flavor, as compared with home, form an additional objection. The Nansemonds, of local growth, are small to medium in size, clear, bright and sound, and sell much higher at all times. The Bermuda is grown here, but objected to by some on account of its size, and the Southern Queen has a limited sale partly on same account. During January and February of this year, the Southern stock ranged from $2.00 to $2.50 per barrel, while the local supply sold at $1.00 a barrel higher, and that difference will continue to exist in values for reasons named. Our market is usually bare of sweet potatoes from the middle of May to middle of August, when the new crop appears.

First receipts from the South appeared about the 25th of July, and sold at $3.00 to $4.00 per barrel, but soon declined $3.00, for home-grown appeared a week later, showing early receipts from Southern points were much later than they should be. First week in August, home-grown are selling at $2.00 per bushel and consignments about $3.50 per barrel for average receipts, which are very light for some cause, and prices show but a small decline up to the 20th of the month, when we find home-grown at $1.50 to $1.75 per bushel, and Southern Yams, $3.00 to $3.25 per barrel; September 1st, home-grown, $4.00 per barrel and shipments $2.25 to $3.00 per barrel; September 15th, home-grown Nansemonds, $3.50 per barrel, Bermuda, $3.00, and Southern shipments, Yams, etc., $1.75 per barrel. It is unnecessary to pursue the course of the market any further. First receipts, small lots, usually appear in boxes, one-half or bushel, and later in barrels.

As already stated, prices were away above those of several preceding years, and last years' figures can't be used as a basis for future operations.

CAULIFLOWER.

It is quite strange, but true, that this very desirable vegetable does not appear among the receipts of goods from the South. In our annual trips throughout the South we have never seen a crop of Cauliflower. We think its general neglect arises from a lack of experience essential to its success in its cultivation. The market here is never crowded with it, except for a short season, when the home-grown supply or season is at its height, and the receipts the largest of the year.

It is an exceedingly valuable and profitable crop in the East, and could be made much more so we think in the South. The Southern growth should be coming here in March and throughout April. There are some (very small heads) in our market from local growers, raised in hot beds, cold frames, etc., during April, and selling at $2.00 to $2.50 per dozen. However, the principal reason that cauliflower is not more generally cultivated in the South is owing to the scarcity of suitable locations. Its natural and most congenial home is by the sea shore, the moisture inseparable from such a locality having a beneficial influence. It has been, in fact, asserted by eminent authority, that it can't be grown, except under such favorable conditions or influences, notably, salt water. This statement, however, will not stand in the face of the fact that this county produces magnificent crops of cauliflower every year, and we are a thousand miles from the coast or salt water, and hence we say that the profits the business in the South can be made to yield are unknown to most vegetable growers, and we would suggest giving the matter a fair trial, to a limited extent at least.

ASPARAGUS

Can be grown profitably for Northern markets, but we cannot encourage any shipments here. Our own gardeners furnish it so extensively and begin so very early in the season, in January, that there is scarcely a profitable opening for outside parties at any time of the year, though the various markets represented in this book can doubtless offer more encouragement. Asparagus is a specialty with many gardeners around here, and they commence shipping to other markets early in the season.

CELERY.

Up to four or five years ago Chicago enjoyed the celery patronage of the West and South, but about this time Kalamazoo, Michigan, embarked in the business, and in no small way either. In the vicinity of that town there was for years a broad expanse of swamp land—nearly two thousand acres—which a native assured us was so poor and desolate the birds would not fly over it, and it could not be sold at any price. One day a practical celery grower came along, secured a portion of this tract, underdrained and prepared it, and in a year was making money very fast—a fact a dozen or more others were not slow to discover, and to-day this immense tract of land is an almost unbroken celery bed—owned and operated by perhaps one hundred different growers and shippers. The business there grew with wonderful rapidity, and shipments found their way to almost every dealer from Maine to Florida—and for nearly six months in the year—July to January, they crowd every market. The whole acreage averages two crops a year—the first growth commencing to move in July and August and the later one October and November—and in a few instances three crops are grown—so rich and appropriate is the land for the purpose.

The business proved a boon to the town, which has nearly doubled in wealth and population as the result—the sales of last year's output being placed at half million dollars. About one thousand eight hundred persons are employed in handling it during the season.

The business at Chicago declined in proportion to the growth of the traffic in Michigan—but through the agency of lower prices a much wider and greater demand grew up—until other points commenced to grow and ship. During the past season Jackson, Michigan, shipped here the finest celery that has ever been offered here—much larger and more attractive stock than the Kalamazoo yield. It was the White Plume variety, comparatively new here, but caught the eye of all purchasers, and at much higher prices than prevailed for that from other points. The celery is white throughout, leaves and all.

The local crop of celery has not been large, but it comes in late—usually December, January and February, and averages, in consequence, good prices.

New Orleans, which heretofore has made but small shipments of celery, raised an enormous crop this past winter, and shipped here quite freely through January and February, and up to this writing March 5th. In size, quality, color, etc., it differs widely from all other receipts. It is a large, tall, rank growth, very poorly bleached, not as crisp or tender as the trade demands, but being without competition— especially during February and March, in this and other markets, finds buyers at good prices, 50c. to 75c. per bunch (twelve stalks).

It is possible that some improvement in quality will follow at New Orleans or other Southern points in the next few years. A better article would find liberal buyers in all the markets of the country during the spring months, when they are bare.

LETTUCE

Has been coming to this market freely during January and February, and up to this writing (March 7th), from New Orleans. It came principally in barrels, a part of which receivers put into bushel boxes on arrival, to accommodate small buyers. The prices ranged from $3.00 to $5.00 per barrel—possibly averaging $4.00. Local growers, too, have been bringing in some almost daily, which sold much higher, being bright, clean and fresh, while consignments do not look so inviting. The New Orleans receipts were mainly sandy and splashed, and none of it really clean. All had to be washed off clean and rehandled before it was offered to the consumers.

We cannot, however, urge any extensive cultivation for this market, because if two or three other points shipped as freely as New Orleans did, and arrived at same time, the market would be glutted, when the demand is so limited and weather such that it can't be exposed for sale at the market stands.

OTHER VEGETABLES.

SPRING ONIONS come in every spring, in all sorts of packages and conditions, and rarely bring much over the express charges. If they are cleaned off nicely, the roots trimed and tied up in bunches of six, no dead or discolored leaves, only the pure white and green colors visible, and packed in crates or boxes, ventilated, they will bring much more than the usual receipts, which are pulled out of the ground and thrown into boxes without any attempt to clean, pack, or render them attractive to buyers, who are accustomed to see all such goods here in the most attractive condition at the various stands and stores where handled.

RADISHES suffer more while in transit than perhaps any other vegetable, and a few crushed leaves will soon start the whole lot on their way to destruction or decay. They should not be shipped with the soil still clinging to them, the way onions sometimes come. Should be washed off clean and tied in bunches, and dry and cool before packed in ventilated packages—not as large as barrels, however, for our market. Must advise going slow for this market.

OKRA OR GUMBO can be grown profitably for our market in a limited way. Mississippi shipped here successfully the past two years. It comes in one-third bushel boxes—beginning at $1.00 per box and soon declining to 50c. It is so light, express charges do not amount to much.

The bulk of the receipts were large, greenish pods—nearly twice as large as the local growth—entirely too large to meet the views of buyers. The dwarf, or smaller-growing varieties, are therefore, most desirable, and should only be cultivated for northern markets.

EGG PLANTS are so heavy, the charges, especially if by express, become a serious matter. They can come here profitably, for a while at least, especially when shipped by fast freight, with melons, cantaloupes, and other goods coming that way from many points. Ship none that are in the least injured or faulty, and only in ventilated barrels or boxes.

SQUASH comes regularly from the south; mainly from New Orleans and Mobile, and usually bring paying, but not big prices. It comes successfully in ventilated boxes or barrels, and wrapped in coarse brown paper; the usual price, according to receipts and circumstances, are from 25 to 50 cents per dozen. Pack none that are speckled, soft, or otherwise injured or faulty. Nearly all come through successfully by fast freight.

We have already reviewed at length the kind of vegetables we recommend for this market. The remainder we bunch together and consider very doubtful, and would advise opening communications with the various other markets represented elsewhere in this work. We frequently have, early in the season, consignments of beets, early corn, carrots, parsnips, turnips, etc., etc., but on an average found only express charges in them, and not always that much. They wilt so while in transit that they rarely reach here in a presentable condition, such as consumers are accustomed to, especially freight receipts and express charges are most too high for them.

The great number of gardeners in the vicinity of this city now accomplish so much through the aid of hot beds, cold frames and early forcing, that quite a list of vegetables can be had here at the stands throughout the year. The changes, improvements and progress developed in this industry here, within the past few years, are simply surprising, and surpass those near any other large city in the West.

Other Markets and Business Cards

Represented in this work are not by any means the least interesting portion of this pamphlet. They form interesting reading and constitute a part of the information which should be attached to a work of this kind, and that is why they are there. Some of these establishments have business all over the Union, and all stand high in their special line. The various interests represented, as well as all the commission houses, embrace a number of firms that the thousands whose hands this little volume will fall into, should do more or less business with every year. We have done business with most of these firms for years, and known them to be thoroughly reliable in all respects, and they are entitled to your confidence and patronage.

Their names could not appear in this volume at any price, unless we knew them to be entirely reliable and trustworthy. We have admitted only such markets and such callings as we know our readers are interested in.

IN CONCLUSION

I will say that this work has been somewhat hurriedly written, for there are no leisure hours for a commission merchant, who is doing a paying business. It is, although an improvement on a similar work written two years ago, not as complete a guide as I desired to write. It covers, however, all the principal points, and will be a great help to new men seeking a livelihood in the business, for it will save them some experimental work and help to avoid the costly errors which the inexperienced are exposed to. I believe, however, from the many approving letters I have received in regard to it, that it is the most practical and instructive work that has yet been given to the public on the subject, and although free to all applicants on receipt of stamps to pay postage, would readily sell at 25 cents per copy. The labor and expense, coupled with the necessary experience essential in getting it out, is such that no other firm in the United States has attempted it or anything similar, and we shall not do so again for three years.

P. M. KIELY.

St. Louis, March 10th, 1888.

CHICAGO.

Established 1865.

A. S. MALTMAN & CO.,

FRUIT AND VEGETABLE

114 & 116 S. WATER STREET,

CHICAGO, ILLS.

Refer, as to Financial Standing or Business Integrity, to any Commercial Agency, Business House or Bank in Chicago.

CONSIGNMENTS OF FRUIT AND VEGETABLES SOLICITED.

☞ Orders for Fruit and Vegetables in car-load lots, or smaller quantities, filled with care and dispatch.

ESTABLISHED 1875.

NEBRASKA COMMISSION HOUSE.

LINCOLN, NEBRASKA.

If you have any Fruits or Produce to ship to this market, give us a trial. We make a specialty in these lines. Quotations, Stencils and Shipping Cards sent promptly on application.

REFERENCES:

WM. T. COLEMAN & Co., San Francisco and Chicago.
A. BOOTH & SON, Chicago and Baltimore.
W. R. STRONG & Co., Sacramento, California.
H. P. STANLEY & SONS, Chicago, Ills.
Any Bank or Wholesale House in Nebraska.

ESTABLISHED 1880.

SOUTH LAWRENCE
Small=Fruit Nursery.

B. F. SMITH,

Box 6.   Lawrence, Kas.

I Make a Specialty of Growing STRAWBERRY PLANTS for Nurserymen and large Planters.

MY COLLECTION EMBRACES 75 VARIETIES,

Including the latest new sorts. I have also on trial, some promising New Seedlings of my own origin.

—o LET ME FIGURE ON YOUR WANTS. o—

A Trial Order will convince you that my Strawberry Plants are second to none.

A PROMINENT FEATURE OF MY PLANT BUSINESS

Is never to send out a two-year old Strawberry Plant on any account, and never to send out Plants not true to name.

FOR SALE THIS YEAR (1888) 800,000 STRAWBERRY PLANTS. PRICE LISTS FREE.

For my 1888 Illustrated Small-Fruit Manual five 2-cent stamps.

REFERENCE:

NATIONAL BANK, Lawrence, Kas.;
G. G. JOHNSON, President, Kansas State Horticultural Society, Lawrence.

Correspondence Solicited.

B. F. SMITH,
Sec'y Douglas Co. Horticultural Society,

Box 66.   **LAWRENCE, KAS.**

BENNETT & HALL

(Established by E. HALL, 1861,)

Commission Merchants,

—AND DEALERS IN—

◁ FRUIT, POULTRY, GAME ▷

And All Kinds of Country Produce,

161 WEST STREET,

Corner Park Place.   **NEW YORK.**

Refer to Publishers of this Book and Irving National Bank, New York.

The Most Extensive and Original Advertisers in the Produce Business.
Have Correspondents in all the Principal Cities and Towns in the United States.
Make Quick Sales and Prompt Returns.
Ship Goods to Order.
Answer Inquiries Promptly.
Attend to Sales Personally.
Have Best of References.

～～～

N. B.—Send for Stencils, Cards and Shipping Directions. Make a Trial Consignment or send an order. Do not-fail to call when you visit the city.

BENNETT & HALL,

Members of the Mercantile Exchange.   *161 West Street, NEW YORK.*

ESTABLISHED 1856.

D. M. Ferry & Co.

Seedsmen,

DETROIT, - MICH.,

Growers, Importers and Dealers in Vegetable, Flower and Field

SEEDS.

We make a specialty of supplying choice Peas, Beans and all other Vegetable Seeds to Truckers and Market Gardeners in all parts of the country.

Our Seed Annual, containing full descriptions, prices and other needed information, mailed free on application.

Ready at Christmas each year. Send for it.

Address,

D. M. FERRY & CO.,

Detroit, Mich.

ESTABLISHED IN 1872.

J. W. SHORT & BRO.,
Successors to J. W. SHORT.
GENERAL COMMISSION MERCHANTS
—— For the Sale of ——
FRUIT, PRODUCE AND VEGETABLES,
Nos. 321 and 323 SECOND STREET,

REFERENCES: { German Insurance Bank. Messrs. Jefferson & Co. } **LOUISVILLE, KY.**

WRITE FOR STENCIL AND INFORMATION.

ESTABLISHED Oct. 17th, 1877.

CHAS. H. GOLDSMITH,
TERRE HAUTE, IND.,
Produce Commission Merchant,

Specially Small Fruits and Vegetables. Oranges, Lemons and Bananas, car load. Late in fall of year, on Potatoes, Apples, Cabbage and Onions. Stencils and information promptly furnished. Prompt returns made. Population this city, forty thousand.

REFERENCES: McKeen & Co., Bank, Adams or Express Agents, this city, or P. M. Kiely & Co.

Truly yours,

CHAS. H. GOLDSMITH.

SEED USUALLY SOWN UPON AN ACRE.

Barley, broadcast........2 to 3 bus.
Beans, Dwarf, in drills..1½ bus.
Beans, Pole, in hills....10 to 12 qts.
Beets, in drills..........5 to 6 lbs.
Broom Corn, in hills.....8 to 10 qts.
Buckwheat...............1 bus.
Cabbage, to transplant...¼ lb.
Carrot, in drills.........3 to 4 lbs.
Chinese Sugar Cane..... 12 qts.
Corn, in hills............8 to 10 qts.
Corn, for soiling........3 bus.
Cucumber, in hills......2 lbs.
Flax, broadcast..........1½ bus.
Hemp....................1½ bus.
Mustard, broadcast......½ bus.
Melon, Musk, in hills...2 to 3 lbs.
Melon, Water, in hills...4 to 5 lbs.
Millet, broadcast........1 bus.
Oats, broadcast..........2 to 3 bus.
Onion, in drills..........5 to 6 lbs.
Onion, for sets, in drills, 30 lbs.
Onion Sets, in drills.....6 to 12 bus.
Parsnips, in drills.......4 to 6 lbs.
Peas, in drills............1½ bus.
Peas, broadcast..........3 bus.
Potatoes, cut tubers .... 10 bus.
Pumpkin, in hills........4 to 6 lbs.
Radish, in drills ....8 to 10 lbs.
Rye, broadcast...........1½ to 2 bus
Sage, in drills............8 to 10 lbs.
Salsify, in drills..........8 to 10 lbs.
Spinach, in drills ....10 to 12 lbs.
Squash, bush. var. in hills, 4 to 6 lbs.
Squash, running " "   3 to 4 lbs.
Tomato, to transplant....¼ lb.
Turnip, in drills..........½ to 2 lbs.
Turnip, broadcast ......3 to 4 lbs.
Vetches, broadcast......2 to 3 bus.
Wheat...................1¼ to 2 bus.

PLANTS AND TREES TO SET TO THE ACRE.

Distance.	Number.	Distance.	Number.
1 foot by 1 foot 43,500		6 feet by 6 feet................1,210
1½ feet by 1½ feet..............19,360		9 " 9 " 537
2 " 2 " 10,890		12 " 12 " 302
2½ " 2½ " 6,970		15 " 15 " 194
3 " 1 " 14,520		18 " 18 " 134
3 " 2 " 7,260		20 " 20 " 105
3 " 3 " 4,840		25 " 25 " 70
4 " 4 " 2,722		30 " 30 " 40
5 " 5 " 1,742		40 " 40 " 37

WEIGHT OF PRODUCE.

	Lbs.		Lbs.
Wheat	60	Blue Grass	14
Corn, Shelled	56	Osage Orange	33
Corn, in the Ear	70	Coal	80
Corn Meal	50	Salt	50
Rye	56	Potatoes, Irish	60
Oats	32	Potatoes, Sweet	50
Flax Seed	56	Onions	57
Buckwheat	52	Turnips	57
Barley	48	White Beans	60
Hungarian Grass	48	Peas	60
Millet	50	Split Peas	60
Clover	60	Castor Beans	46
Hemp	44	Green Apples	50
Malt	34	Dried Apples	24
Timothy	45	Onion Top Sets	28
Sorghum Cane	50	Dried Peaches	33
Red Top Grass	14	Bran	20
Orchard Grass	14	Peanuts, Dry Southern	22

GRASS SEED TO THE ACRE.

White Clover........3 to 5 pounds.	Hungarian Grass....1 bushel.
Red Clover..........10 to 15 pounds.	Blue Grass..........1½ to 3 bushels.
Lucerne Clover.....6 to 8 pounds.	Rye Grass..........1½ to 2 bushels.
Alsike Clover......4 to 6 pounds.	Orchard Grass .....1½ to 2 bushels.
Timothy............12 pounds.	

W. P. MESLER & CO.,

Manufacturers and Dealers in all kinds of

Fruit and Vegetable Packages.

FACTORY AT

COBDEN, Union County, -- -- ILLINOIS.

42 miles North of Cairo on I. C. R. R., and three miles from St. Louis & Cairo R. R.

We keep on hand large quantities of Hallock and Leslie Quarts and Crates, ⅓ bushel and bushel boxes, and can fill orders early with dry material; saving consumers considerable on freight.

We also keep on hand Wire Nails, Tacks, Tack Hammers, Forms for making Quarts, Wire and Wire Sewing Box Machines for sewing the boxes together.

Refer to PARKER EARLE, Pres. Miss. Val. Hort. Society.
J. H. & H. E, McKAY, Madison, Miss.
W. M. SAMUELS, Clinton, Ky.
P. M. KIELY & Co., St. Louis.

SEND FOR PRICE LIST.

FARMERS!
Feed your Land and it will Feed you.

Use MAYER'S
Anchor Brand "Fertilizers"

— ON ALL —

FIELD AND GARDEN CROPS,

Thereby increasing the yield 50 to 100 per cent., and maturing the crops much earlier.

SEND FOR OUR

✵ MEMORANDUM ✦ POCKET ✦ BOOK, ✵

Giving full directions, etc.

A. B. MAYER M'F'G CO.,
ANCHOR BONE WORKS AND ANCHOR FERTILIZER WORKS,
ST. LOUIS, MO.

We also make a brand especially adapted for Oranges, Florida crops, called the "Tankage Fertilizer."

ESTABLISHED 1876.

B. M. TANNER,
Commission Merchant,
521 WALNUT STREET,
KANSAS CITY, MO.

Special Attention Given to Early Fruits and Vegetables in Their Season.

NOTICE TO SHIPPERS.—Kansas City has grown so rapidly during the past few years that it is now one of the great and profitable shipping points of the growing West. This city has a population of two hundred thousand people, and in addition, a very large floating population, which daily consumes a vast quantity of the early products of the South. Kansas City is undoubtedly the best distributing point in the whole West, having within a radius of sixty miles no less than eight cities, ranging in population from eight to forty thousand, which very largely draw their supplies from this market.

Shipping Stencils and all desired information furnished upon application.

REFERENCES:
GERMAN AMERICAN BANK, this city.
BRADSTREET'S or DUN'S MERCANTILE AGENCIES.
P. M. KIELY & Co., St Louis.
CHAS. H. SCHENCK, New Orleans.

ESTABLISHED IN 1880.

—THE—
G. G. LIEBHARDT COMMISSION CO.
IMPORTERS AND GROWERS,
AGENTS FOR
FOREIGN AND DOMESTIC FRUITS,
1624-1630 Holladay St. -- DENVER, COL.

✠ SPECIALTIES: ✠

Apples, Oranges, Lemons, Cranberries, Bananas, California Green Fruit, Melons, Grapes, Strawberries and Sweet Potatoes.

→WE are the only house west of St. Louis that can handle our specialties with same promptness as they are handled in large Eastern markets, and the only house that can close out a car of Strawberries to advantage on arrival.

CORRESPONDENCE SOLICITED.

PRIVATE TELEGRAPHIC CIPHER CODE, Stencil Plate's, Price Currents, etc., free on application.

WILLIAM B. CURTH,
Produce ✢ Commission ✢ Merchant
AND WHOLESALE DEALER IN
Foreign and Domestic Fruits, Butter, Eggs and Cheese,

807 Water Street, - - SANDUSKY, OHIO.

I solicit consignments of Water Melons, Cantaloupes, Berries, Vegetables of all kinds, also all kinds of Dried and Fresh Fruit, Potatoes, Onions, Nuts. Butter, Eggs, Cheese, Poultry, etc. Send for Stencil and Market Quotations.

Yours respectfully,

WILLIAM B. CURTH.

Colman's Rural World.

ESTABLISHED BY		CONDUCTED BY
NORMAN J. COLMAN.		**CHALMER D. COLMAN.**

PUBLISHED WEEKLY.
ONLY ONE DOLLAR A YEAR.

The RURAL WORLD is the oldest Agricultural and Horticultural Journal in the Mississippi Valley, and, upon comparison, will be found to be equal to the best. Send for a sample copy and see.

It is the leading Horticultural Journal of the West, and publishes the cream of the current news in that department.

For the stock raiser, horse breeder, wool grower and the general farmer, it publishes more and better information than any other paper. Sample copies free. Address,

C. D. COLMAN,

705 Olive Street - - - *ST. LOUIS, MO*

SUTLIFF BROS.
GENERAL COMMISSION MERCHANTS
CEDAR RAPIDS, IOWA.

SPECIAL ATTENTION GIVEN TO

SOUTHERN FRUITS AND VEGETABLES
FOR WHICH WE HAVE A LONG SEASON HERE.

Correspondence Solicited.

W. S. McMAINS & CO.,
Produce Commission Merchants,
—— And Wholesale Dealers in ——

Fruits, Berries, Early Vegetables, Melons, Eggs, Butter Hides, Etc., Etc.,

403 Walnut Street, - - KANSAS CITY, MO.

REFERENCES { National Bank of Commerce, Kansas City, Mo.
BY { Pacific Express Company.
PERMISSION : { Commercial Agencies.

We solicit your consignments of Berries, Fruits, Water-Melons, Cantaloupes, New Potatoes and Tomatoes, and all other Vegetables, Etc. Send for Stencil and Market Quotations.

Yours, very respectfully,

W. S. McMAINS & CO.

Established in 1866, 120 S. WATER STREET.

Established in 1886, 159 S. WATER STREET.

BARNETT BROS.,
CHICAGO,
COMMISSION and FRUIT DEALERS.

HAVING been long in the business, we offer our services to FRUIT SHIPPERS, and are fully satisfied we can serve you to advantage. Since the last issue of this book, we have been compelled to seek new quarters, and now have the whole of 159 South Water Street, 28x150, four stories high, well lighted and equipped with a full corps of trained employes, so that the personal advantages we offer are unsurpassed by any house in the business. The advantages of Chicago, as a market, are somewhat known to the fruit grower. We do not urge shipments to our market *regardless of price*, but we shall be pleased to answer correspondence at all times, and urge on all intending to ship, to obtain reliable information about this market before shipping, if possible.

Stencils will be furnished free, and we will do all we can for the best interests of the shipper in the sale of **Fruits and Vegetables.**

We refer, by permission, to the Agents of the SOUTHERN, ADAMS', AMERICAN and UNITED STATES EXPRESS COMPANIES; also to P. M. KIELY & Co.

BARNETT BROS.,
159 South Water Street, - - - - CHICAGO.

Agents for the FLORIDA FRUIT EXCHANGE.

www.ingramcontent.com/pod-product-compliance
Lightning Source LLC
Chambersburg PA
CBHW020151170426
43199CB00010B/993